BIOINFORMATICS:
A Biologist's Guide to Biocomputing and the Internet

Other BioTechniques® Books Titles

Gene Cloning and Analysis by RT-PCR
P.D. Siebert and J.W. Larrick (Eds.)

Apoptosis Detection and Assay Methods
L. Zhu and J. Chun (Eds.)

Protein Staining and Identification Techniques
R.C. Allen and B. Budowle

Immunological Reagents and Solutions: A Laboratory Handbook
B.B. Damaj

Affinity and Immunoaffinity Purification Techniques
T.M. Phillips and B.F. Dickens

Yeast Hybrid Technologies
L. Zhu and G. Hannon (Eds.)

Viral Vectors: Basic Science and Gene Therapy
A. Cid-Arregui and A. García Carrancá (Eds.)

Ribozyme Biochemistry and Biotechnology
G. Krupp and R.K. Gaur (Eds.)

Antigen Retrieval Techniques: Immunohistochemistry and Molecular Morphology
S.-R. Shi, J. Gu, and C.R. Taylor (Eds.)

SNP and Microsatellite Genotyping: Markers for Genetic Analysis
A. Hajeer, J. Worthington, and S. John (Eds.)

Gene Transfer Methods: Introducing DNA into Living Cells and Organisms
P.A. Norton and L.F. Steel (Eds.)

BIOINFORMATICS:
A Biologist's Guide to Biocomputing and the Internet

Stuart M. Brown
NYU Medical Center
Department of Cell Biology and
Research Computing Resource
New York, NY, USA

A BioTechniques® Books Publication
Eaton Publishing

Stuart M. Brown, PhD
NYU Medical Center
Department of Cell Biology and
Research Computing Resource
550 First Avenue
New York, NY, 10016

Library of Congress Cataloging-in-Publication Data

Brown, Stuart M., 1962-
Bioinformatics : a biologist's guide to biocomputing and the internet / Stuart M. Brown.
p. cm.
Includes bibliographical references and index.
ISBN 1-881299-18-X
1. Bioinformatics. 2. Molecular biology--Data processing. I. Title.

QH506 .B767 2000
570'.285--dc21

00-046247

ISBN 1-881299-18-X

Printed in the United States of America

9 8 7 6 5 4 3 2 1

Eaton Publishing
BioTechniques Books Division
154 E. Central Street
Natick, MA 01760
www.BioTechniques.com

Francis W. Eaton: *Publisher and President*
Stephen Weaver: *Director and Editor-in-Chief*
Christine McAndrews: *Managing Editor*
Sandy Lamont: *Production Manager*
Peggy Hewitt: *Cover Designer*

FOREWORD

Two major events in the 20th century—the discovery of the structure of DNA and the development of methods for rapidly determining the nucleotide sequence of DNA—have led to an information explosion in the biological sciences. More recently, the worldwide efforts to sequence the entire genome of many model prokaryotic and eukaryotic organisms have resulted in the international Human Genome Project. This information now forms the basis for the anticipated 21st century research in understanding heredity at the chemical level and will result in a deeper insight into the structure and functions of the genes encoded in this heredity material. Today, scientists interested in biological systems have at their disposal a massive amount of ever-growing DNA sequence data generated by genome sequencing centers, as well as the additional biological information being discovered in parallel. Therefore, to remain current with this information, researchers in many disciplines not only must understand this whole genome data within the context of our own research, but also must efficiently apply this data to elucidate the specific function(s) of these regions and the gene products encoded therein.

Just when these problems seem overwhelming, Stuart Brown provides us with a road map leading to the computational tools necessary to make an educated decision about which of the computer-based approaches can solve our specific problems and describes them and their use in understandable terms. *Bioinformatics* opens with an overview of the Human Genome Project, its associated data, and the computer tools needed to optimally utilize this newly acquired genomic sequence information in real time. Once the latest data of interest are captured, he leads us through the complex maze of programs that allow comparing the regions of interest to related sequences by multiple sequence alignment, locating the genes and regulatory regions, and studying the structure of the encoded protein and unique features. Of course, once predicted, these features need experimental confirmation. Therefore, Dr. Brown also provides a detailed discussion of the computer-based tools that can facilitate these lab-based studies. From restriction mapping and subsequent gene construction, to primer design for PCR and DNA sequencing, to a detailed discussion of the available DNA sequencing related software, Dr. Brown carefully and clearly illuminates the entire process. Once similar genes from many species have been analyzed individually, a detailed comparative genomics via taxonomy and phylogenetic studies is in order; here, Dr. Brown provides a discussion of the multitude of available computer programs that can help researchers know which analysis is more appropriate and how the results can be interpreted. Finally, through a detailed listing of the presently available bioinformatics software as well as a thorough glossary, Dr. Brown brings the practicing biologist, biochemist, or molecular biologist to a working introduction into the complex arena of bioinformatics.

Bruce A. Roe, PhD
The University of Oklahoma
October 2000

PREFACE

It is nearly impossible to function as a molecular biologist today without some basic skills in bioinformatics. Every laboratory technique from simple cloning and PCR to gene expression analysis requires the manipulation of data on a computer. Biologists need a basic understanding of the algorithms that underlie the DNA and protein analysis tools that are used for these essential research tasks.

This book provides the biologist, student, or interested amateur a practical overview of the field of bioinformatics. Subjects include searching sequence databases for genes by name, similarity (i.e., "homology") searching, restriction mapping and PCR primer design, pairwise and multiple alignment, locating promoters in DNA and functional motifs in proteins, DNA sequencing, and phylogenetics (evolutionary analysis). The basic logic behind each type of analysis is explained without overwhelming the reader with the details of the mathematics that are used. Bioinformatics programs for mainframes and personal computers, and those freely available on the Web, are discussed and evaluated.

Stuart M. Brown
August 2000

Contents

Section I: An Overview of Bioinformatics

Section II: Computer Tools for Sequence Analysis

Section I

An Overview of Bioinformatics

1 Introduction

A BIOINFORMATICS SURVIVAL GUIDE

Bioinformatics can be defined as the use of computers for the acquisition, management, and analysis of biological information. It exists at the intersection of molecular biology, computational biology, clinical medicine, database computing, the Internet, and sequence analysis. The key element that these definitions share is information management. Driven by automated DNA sequencing technology and the Human Genome project, analysis of DNA and protein sequence data has spurred the dramatic growth of bioinformatics in the 1990s. New high-throughput genomics technologies such as DNA microarrays are poised to create yet another revolution in bioinformatics in the next decade.

This book provides the biologist, student, or interested amateur a practical overview of the field of bioinformatics. The goal is to enable researchers to evaluate and choose the appropriate software, databases, and/or Web sites for various analysis tasks and to intelligently select options within various software packages. This is the insider's guide to the arcane lingo of "ktups" and "gap extension penalties" used in bioinformatics programs.

Bioinformatics is a new scientific discipline created from the interaction of biology and computers. As such, it is changing and developing rapidly, and it is therefore rather difficult to define. The term "informatics" has been in general use for some time in both the computer and health care industries to denote all aspects of information technology and information management. Medical informatics is focused on computerized medical records and billing systems, but also includes databases of medical literature and decision support software.

The most important trends currently affecting bioinformatics can be summarized in three terms: sequence data, genomics, and the Internet. Another trend is the scaling up of biological research programs, particularly in the pharmaceutical industry, as well as some University and Government programs such as the Human Genome Project. Scaling up has led to a more automated approach to projects such as DNA sequencing, high-throughput screening of compounds, combinatorial libraries of molecules, and biorational drug design. These types of large-scale

Bioinformatics
By Stuart M. Brown

research programs require a much greater ratio of computational support per laboratory technician as compared to the traditional small laboratory group. Less obvious are the gradual cultural changes that have occurred in biology over the past decade, the ubiquitous growth of molecular biology as a tool that provides insights into every subdiscipline of biology, and the increasing acceptance of computer sequence analysis as a routine laboratory tool for the molecular biologist.

While there have been many incremental improvements, the basic set of bioinformatics sequence analysis tools has not changed much in the past 10 to 15 years. There are *(i)* tools that allow the researcher to locate sequences in public databases by accession number or keyword searching; *(ii)* tools for aligning two sequences or multiple sequences; *(iii)* tools for finding sequences in a database similar to a query sequence (similarity searching); *(iv)* tools to aid in the DNA sequencing process by assembling short overlapping fragments into consensus sequences; *(v)* tools for analyzing the physical and chemical properties of proteins; *(vi)* pattern analysis tools to locate restriction sites, promoters, DNA binding domains, and protein motifs; and *(vii)* tools for phylogenetic (evolutionary) analysis.

The most significant change in bioinformatics in the past decade has been the tremendous amounts of new sequence data that has become available in public databanks, primarily GenBank®. The sequences in GenBank have doubled every 18 months over the past 10 years. With the rapid progress in the sequencing of the genomes of humans, microorganisms, and other model species, it is now possible to perform sophisticated genetic experiments in the computer (*in silico*) either in preparation for, or in place of, laboratory work. Access to complete genomes of different species is enabling new types of genome-level comparative analysis and systematic approaches to bioinformatics such as metabolic reconstruction.

Genomics, the study of biology at the genome-wide scale, is completely changing the Life Sciences and opening remarkable new possibilities for understanding diseases, fundamental biology, and for medical and bioengineering applications. Genomics is based on automated, high-throughput methods for generating experimental data, ranging from gene sequences to gene expression levels (DNA microarray chips), to protein–protein interactions. Genomics technologies are making it possible to study biology as a system, rather than as a disconnected series of reductionist experiments.

It is important to consider the interdependence of genomics and bioinformatics. Genomics produces high-throughput, high-quality data, and bioinformatics provides the analysis and interpretation of these massive data sets. Neither could produce results without the other, yet they traditionally are separated by barriers of methodology and expertise. Genomics is based on the experimental and engineering aspects of molecular biology, while bioinformatics is based on computer science, mathematics, and theoretical approaches to the data.

The rapid advancement of genomics technologies is creating an ever greater integration of computing into the daily life of biomedical researchers. Bioinformatics is coming to occupy a central role in modern biology, integrating data from fields as diverse as enzymology, genetics, structural biology, medical and animal models of disease, etc. The important relationships between these data are being elucidated through their gene associations, and bioinformatics provides the tools for making these associations.

The Internet is pervasive in all bioinformatics work. The Web browser has become the primary means of access to bioinformatics data and programs for most biologists. Primary databases of DNA and protein sequences and genetic maps, as well as derived databases created by the analysis and/or annotation of this information, are available over the Web (this is often the only means of accessing these databases). It is simply not practical to maintain local copies of complete, up-to-date sequence databases on a personal computer (PC). New analytical software is now generally available to be used on Web servers as well as provided as programs to be run on users' own computers. Existing molecular biology software for PCs is increasingly relying on Internet access to databases and remote processing, rather than relying on data and processing power within the PC. More complex mainframe-based molecular biology programs and database systems are increasingly relying on Web interfaces rather than traditional text-based user interaction or custom client software.

THE INTEGRATION OF SOFTWARE, DATABASES, AND THE WEB

A critical failure of current bioinformatics tools is the lack of a single software package that can perform all of the functions that a researcher requires. These functions can be divided into three main categories: *(i)* database searching, *(ii)* sequence analysis, and *(iii)* the display of analysis results in a publishable format. Experienced bioinformatics researchers constantly seek the best computer tools for each type of analysis, which forces them to use a hodgepodge collection of Web servers, mainframe computer programs, and programs running on PCs. The challenge for bioinformatics is to create a single customizable interface that can integrate this constantly changing collection of tools into a coherent work environment. The Web is emerging as this universal interface.

A number of companies are developing mainframe molecular biology computer programs that can be accessed directly from a desktop PC or Macintosh® computer using a Web browser. Pangea Systems (Oakland, CA, USA) has developed a Web-based relational database system know as GeneWorld™ to manage the daily workflow operations of a sequencing laboratory and to automate standard analyses of new sequences, including algorithms for gene finding, domain finding, pairwise and multiple sequence alignment, secondary structure prediction, and functional identification. In October of 1999, Pangea Systems launched a free public version of the GeneWorld software called DoubleTwist™ available at the **http://www.doubletwist.com** Web site.

The popular mainframe Wisconsin Sequence Analysis Package (also know as GCG), marketed and supported by Oxford Molecular Group (Oxford, England,

UK), is also being adapted to the Web. A preliminary version of a Web interface to GCG called Seqweb was released in early 1999. An update to GCG in 2000, to be called SeqStore, promises to provide additional Web features including integration with Oracle® databases (Oracle, Redwood Shores, CA, USA) and automated analysis of sequences with any of the GCG programs.

Molecular Simulations (San Diego, CA, USA) has developed GeneExplorer™, which provides an intuitive graphical user interface and Web-based access to bioinformatics programs and databases. GeneExplorer is designed to determine protein 3-D structure through homology modeling (similarity to known structures). GeneExplorer also includes modules for restriction analysis, virtual mutagenesis, as well as DNA and protein similarity searching.

There are several projects underway to build a complete molecular biology computing package as a free Web interface. The Computational Biology Group at the National Center for Supercomputing Applications (NCSA), University of Illinois, has developed a free Web-based program called The Biology Workbench written in the cross-platform Java™ language (**http://biology.ncsa.uiuc.edu/**). The WorkBench integrates many standard protein and nucleic acid sequence databases and a wide variety of sequence analysis programs into a single interface. The WorkBench allows users to perform database searches, multiple alignments, phylogenetic analysis, protein secondary structure prediction, and motif searches; all without worrying about file formats or command line options. Sequences, projects, and results are saved in a password-protected area for each user, and encryption is used to transmit data between the NCSA computer and the Web browser on the user's computer.

CuraGen (New Haven, CT, USA) has created a suite of genomics tools called CuraTools™, which are available for free on a Web site called the GeneScape Portal™ (**http://curatools.curagen.com**). CuraTools is an integrated suite of bioinformatics tools that allows users to easily perform analyses on collections of DNA and protein sequences. Users can upload their own sequence data that can be edited and stored indefinitely in a password-protected personal file storage area on the CuraGen Web site. Using a simple Web interface, users can rapidly analyze their data and associate it with many other forms of genomic data, following similarity searches, motif analyses, structure–function predictions, etc. Results from the various analyses are displayed with the aid of specially tailored viewers and interpreters. Results can be saved together with personal data sets and search results.

CuraTools include similarity searches, multiple alignment, PCR primer design, DNA sequence assembly, DNA restriction analysis and open reading frame (ORF) prediction, protein analysis including amino acid composition, molecular weight, pI, proteolytic digestion, hydropathy and antigenicity, secondary structure prediction, identification of protein domains, and phylogenetic analysis. Most of these tools are implemented by sending queries out to a wide variety of other Web servers and capturing the results, but this is done seamlessly for the user, and all results are reformatted into a consistent "report" format.

There are several other Web sites that provide a single point of access to aggregations of Web molecular biology tools. They are slightly more useful than simple collections of links such as the famous Pedro's Biomolecular Research Tools (**http://www.public.iastate.edu/~pedro/research_tools.html**). However, these sites do not allow users to save their own sequences, nor do they reformat the results as they are returned from multiple search engines across the Web.

- The BCM Search Launcher (Baylor College of Medicine)
  http://www.hgsc.bcm.tmc.edu/SearchLauncher
- The Biologists Search Palette
  http://www.biozentrum.uni-wuerzburg.de/biolinks/search.html
- Emmanuel Skoufos' Gene Discovery Page
  http://bioinformatics.weizmann.ac.il/gdp/gdp.html

An alternate approach to building an integrated bioinformatics platform is to integrate stand-alone molecular biology programs for the PC with services that are available over the Internet. Vector NTI™ by InforMax (Bethesda, MD, USA) contains an integrated programming language that passes Web addresses and data to a Web browser running on the user's computer. In this way, Vector NTI serves as a launching point for analysis that can be performed on any Web site. Internet access to the ENTREZ and BLAST services from the National Center for Biotechnology Information (NCBI) has been built into the commercial molecular biology programs MacVector™ and OMIGA™ produced by Oxford Molecular Group. The OMIGA program will also be able to connect with GCG running on a mainframe via an interface called GCGLink for database searching and other computationally intensive programs. These programs have essentially built in some Web browser–network computing functionality into their proprietary program interface.

The incorporation of some Internet-based services into desktop computer programs may be a useful solution at the moment, but it has some significant disadvantages when compared with a purely Web-based solution. First, the scientist is forced to work from a single computer where the custom program is installed, rather than being free to work from any computer equipped with a Web browser. Second, either versions of every program must be made available for the variety of different operating systems used in desktop computers (DOS, Windows®, Windows 95, Windows 98, Macintosh, PowerMacintosh, UNIX®, LINUX, SOLARIS, etc.) or else the user will be limited in the choice of computers or in the choice of software available for a particular computer system. Third, there is the continual need for upgrades to the software on each desktop machine as new features are introduced or issues arise of compatibility with new operating systems or other software. In a large institution, it is much more efficient to administer a single mainframe software package accessed by many users over the Web than to have many copies of a program installed on a variety of PCs.

Ultimately, it is likely that the advantages of the Web in providing universal access from any location and any type of computer will prevail over software designed for the individual desktop computer. This new type of Web-based bioinformatics package should seamlessly integrate access to all relevant databases throughout the world (both public and private), perform computationally intensive tasks on mainframe computers, and make use of JAVA programs to create a responsive and graphically complex interface on the user's desktop machine.

INTEGRATED DATABASES

Another major failure of current bioinformatics systems is the inability to access and interlink all relevant data. Significant progress has been made in collecting data from all of the various DNA and protein sequence databases into a few central sources in a consistent format and making this data available via the Internet. By providing a Web server as a universal access point, curators of many diverse databases can provide very powerful computers to perform searches for users using custom query languages designed to cope with the unique features of each database and provide access to a wide variety of content beyond sequences that might include genetic maps, 2-D protein gels, 3-D images of molecules, etc. For example, The ACEDB genome databases maintained by the US Department of Agriculture (USDA) (**http://probt.nalusda.gov:8000/alldbs.html**) contain an extremely wide variety of different data types (maps, genes, alleles, sequences, clones, contigs, probes, germplasm, images, journal citations, etc.). While not perfectly intuitive, the ACEDB format does allow great flexibility in adding new data types for specific databases, while providing a consistent query method and hyperlinking between related database elements

However, each of the derived molecular biology databases contains a different set of "value added" information types, has a different search interface, stores data in a different file format, and uses a different query language. Furthermore, a query in one of the primary sequence databases will not yield information stored in the other databases, nor will it alert the user to the existence of other data relevant to the search. In order to access all of the relevant data, a researcher must know how to search many different databases, each holding different types of data, updated on different schedules, and using different search tools with different interfaces.

It is difficult for a biologist to know what resources to use for a particular type of analysis, and it is confusing to move among different Web servers, each with a different look and feel to its interface. A key challenge for bioinformatics practitioners is to simplify and integrate these diverse databases and Web-based bioinformatics programs, so that the scientific user can go to a single location and use a single search language to find all relevant information for any type of database query. Ideally a single Web interface should intelligently direct a query to all relevant databases or programs (running on servers throughout the world) and then collect the results and display them in a consistent format so that data analysis can flow smoothly from one tool to another.

Some species-specific databases such as the Human Genome DataBase (GDB) and the Saccharomyces Genome Database (SGD) now include links back to the primary DNA and protein databases (GenBank and SWISS-PROT). The BCM Search Launcher Web site (**http://kiwi.imgen.bcm.tmc.edu:8088/search-launcher/launcher.html**) is one approach to an integrated bioinformatics interface. This bare-bones Web site provides access to a tremendous number of powerful Internet-based tools and displays the results of the analyses on a Web page. However, the use of the results of one analysis as the input for another is still a significant challenge for the researcher.

The ENTREZ/PUBMED database at the NCBI has made the most progress in integrating a variety of data types and making it all freely available over the Internet (**http://www.ncbi.nlm.nih.gov/Entrez/**). ENTREZ includes all of the DNA sequences in GenBank, all of the protein sequences in GenPept, cross references to all MEDLINE (PubMed) journal articles, and links to 3-D structure information from the Brookhaven Protein Data Bank (PDB). Each entry in any of these databases is linked to any related entries in any of the other databases. In addition to these direct cross-reference links, one of ENTREZ's most powerful functions is provided by its computation of "neighbors" within each of the databases. The neighbors of a sequence are its homologues, as identified by a significant similarity score using the BLAST algorithm. The neighbors of a PubMed citation are articles that use similar keyword terms in their title and abstract. Structural neighbor information in ENTREZ is based on a direct comparison of 3-D structure using the Vector Alignment Search Tool (VAST) algorithm.

Another example of database integration is the integrated protein database including SWISS-PROT, PROSITE, SWISS-2DPAGE, and SWISS-3DIMAGE, accessible through the ExPASy Web server at **http://expasy.hcuge.ch/**.

In the SWISS-2DPAGE database, there are *Escherichia coli*, *Saccharomyces cerevisiae*, and twelve human tissue 2-D reference maps. To access text information about an identified protein, the user simply selects a spot on the gel image. A hypertext link can then be used to obtain the full SWISS-PROT entry for that protein, displaying protein sequence, domain structure, information on known posttranslational processing and modifications, and references. From SWISS-PROT, the user can select a link to SWISS-3DIMAGE to see the 3-D structure of the protein, if it is known, or to submit the sequence to the SWISS-MODEL 3-D modeling tool. Also, from SWISS-PROT, the user can select links to pertinent information from DNA sequence databases (European Molecular Biology Laboratory [EMBL], GenBank), chromosomal and genomic maps (GDB), bibliographic references and abstracts (MEDLINE), and databases on the association of human proteins with diseases (Online Mendelian Inheritance in Man [OMIM]).

2 Genomics and the Genome Project

GENOMICS

The latest buzzword in the biological research community is "genomics." Everyone is talking about how it will revolutionize biology and medicine and trying to find ways to wedge it into their latest grant proposal. What is it all about, how will it affect us as researchers, and what are the direct implications for bioinformatics?

An operational definition of genomics might be: "The application of high-throughput automated technologies to molecular biology." Genomics technologies include automated DNA sequencing, expressed sequence tag (EST) sequencing, DNA chip microarrays (for the measurement of gene expression and for the characterization of sequence polymorphisms), and various tools for the large-scale analysis of protein expression and function (proteomics). However, it is impossible to separate genomics laboratory technologies from the computational tools required for data analysis. Genomics produces high-throughput, high-quality data, and bioinformatics provides the analysis and interpretation of these massive data sets.

The scientific implications of genomics technology have already proven to be extremely wide ranging. A more philosophical definition of the field would be: "A holistic or systems approach to the study of information flow within a cell." The essential distinction between genomics and classical molecular biology is the consideration of the entire complement of DNA and RNA in a cell or organism as opposed to the classic reductionist approach that looks at the role of just one gene at a time.

GENOME SEQUENCING

The first high-throughput genomics technology was automated DNA sequencing. In the early 1990's, automated DNA sequencing machines allowed individual scientists (and then technicians) to increase their output of DNA sequences from a few thousand base pairs per week up to millions of base pairs per week, with much less effort and greater accuracy and reproducibility. This veritable "fire

Bioinformatics
By Stuart M. Brown

hose" of DNA sequence data created a bottleneck for sequence analysis and annotation—both within the individual laboratories doing the sequencing and at the organizations maintaining sequence databases. This led to a revolution in bioinformatics as automated systems were developed for assembling raw sequences as they came off of the sequencing machines, analyzing and annotating them, and incorporating them into databases. This created ripple effects throughout the scientific community as researchers had to adapt their sequence analysis tools to much larger and more rapidly updated databases.

Even before automated DNA sequencers were widely available, a number of visionary scientists realized that improvements in DNA sequencing technology would soon make it possible to sequence the entire human genome. In 1986, the U.S. Department of Energy (DOE) announced a Human Genome Initiative, which was initially conceived as a 15-year program to map and sequence the human genome. In 1988, the U.S. National Institutes of Health (NIH) created its Office for Human Genome Research led by James D. Watson. In 1989, this Office was elevated to the status of an official NIH institute: The National Center for Human Genome Research (NCHGR). This Institute was later renamed the National Human Genome Research Institute (NHGRI).

In October of 1990, the DOE and NIH formally began joint funding for the Human Genome Project with an initial budget projection of $200 million per year for approximately 15 years in order to complete the sequence of the human genome by the year 2005. The initial 5-year plan (8) for the project set the following goals for 1995 that focused on technology development, mapping, and preparation for large-scale sequencing:

1. Complete a fully connected human genetic map of all chromosomes with sequence-tagged site (STS) markers spaced an average of 2 to 5 centimorgans apart (approximately 100 000 bp).
2. Generate overlapping sets of cloned DNA (contigs) with continuity over lengths of 2 million base pairs for large parts of the human genome.
3. Improve current methods and/or develop new methods for DNA sequencing that will allow large-scale sequencing of DNA at a cost of 50 cents per base pair.
4. Determine the sequence of an aggregate of 10 million base pairs of human DNA in large, continuous stretches.
5. Sequence an aggregate of approximately 20 million base pairs of DNA from a variety of model organisms, focusing on stretches that are 1 million base pairs long.

In 1993, it was announced that the original goals of the Human Genome Project would be met on or ahead of schedule. A revised set of goals was developed that included sequencing an aggregate of 80 million base pairs of DNA by 1998 and finishing the complete sequences of the *Escherichia coli* and *Saccharomyces cerevisiae* (yeast) genomes by 1998 or earlier (3). It is interesting to note that although the original 1990 plan called for funding of $200 million per year, only

$170 million per year was actually spent from 1990 to 1993, so the project was already running under budget and ahead of schedule.

Dr. J. Craig Venter, president and director of The Institute for Genomic Research (TIGR, Rockville, MD, USA), announced in May of 1998 a joint venture in collaboration with Perkin-Elmer (the dominant manufacturer of automated DNA sequencing equipment) to create a new company called Celera Genomics (Rockville, MD, USA), which would operate a sequencing facility with a capacity greater than that of the previous combined world output. Applying new sequencing technologies being developed by Perkin-Elmer's Applied Biosystems Division to sequencing strategies pioneered by Dr. Venter and others at TIGR, Celera Genomics would complete the sequencing of the human genome by December 31, 2001. As a proof of concept, Celera Genomics would first complete the sequencing of the *Drosophila melanogaster* genome. The new company would also build the scientific expertise and bioinformatics tools necessary to extract valuable biological knowledge from genomic data, including the discovery of new genes, development of polymorphism assay systems, and databases for the scientific community.

Inspired by Celera Genomics, the DOE-NIH Human Genome Project announced an updated set of goals in 1998 (4):

1. Finish the complete human genome sequence by the end of 2003.
2. Finish one-third of the human DNA sequence by the end of 2001.
3. Achieve coverage of at least 90% of the genome in a working draft based on mapped clones by the end of 2001.
4. Complete the sequence of the *Caenorhabditis elegans* (roundworm) genome by 1998.
5. Complete the sequence of the *Drosophila* (fruitfly) genome by 2002.
6. Generate sets of full-length cDNA clones and sequences that represent human genes and model organisms.
7. Create a human single nucleotide polymorphism (SNP) map of at least 100 000 markers.
8. Identify common variants in the coding regions of the majority of identified genes.
9. Develop technologies for rapid, large-scale identification and/or scoring of SNPs and other DNA sequence variants.

In September of 1999, Celera Genomics announced that it had completed the sequencing of the *Drosophila* genome and would now devote its full sequencing efforts to the human genome. The DOE-NIH Human Genome Project matched that announcement in October of 1999 with a plan to publish its own complete "draft genome sequence" by the spring of 2000 (11). The Genome Project is probably the world's only global collaborative project that keeps reducing its projected budget and moving up its projected completion date. You can read the history of the project for yourself at the Human Genome Project Information Web site: **http://www.ornl.gov/hgmis/project/hgp.html**.

RAW GENOME SEQUENCE DATA

If you want to get preliminary human genomic sequences before they have been fully completed, all of the sequencing centers participating in the International Human Genome Project contribute their latest data to the NCBI's **htgs** database (high throughput genome sequencing): **http://www.ncbi.nlm.nih.gov/HTGS**.

There is also a special BLAST web page to compare your query sequence against **htgs** data: **http://www.ncbi.nlm.nih.gov/genome/seq/page.cgi?F=HsBlast.html&&ORG=Hs**.

The NCBI has complete genome sequences for over 600 organisms (as of 9/2000) including 29 bacteria, 7 Archaea, and 3 eukaryotes (*C. elegans*, *Drosophila*, and yeast), as well as almost 600 viruses (**http://www.ncbi.nlm.nih.gov:80/PMGifs/Genomes/org.html**). The NCBI also maintains a web page which allows BLAST searching of all the completed genomes as well as many partially sequenced genomes: **http://www.ncbi.nlm.nih.gov/Microb_blast/unfinishedgenome.html**.

However, the NCBI microbial genomes site does not contain the most up-to-the-minute data on microbial genomes. For example, the NCBI has data from *Actinobacillus actinomycetemcomitans* that was updated on July 2, 1998, but the University of Oklahoma's Advanced Center for Genome Technology Web site (**http://www.genome.ou.edu/**) has *A. actinomycetemcomitans* data in the form of contigs from current sequence reads that are updated nightly.

The NHGRI (National Human Genome Research Institute) has a list of links on its Web site 9 (**http://www.nhgri.nih.gov/Data**) to the major genome sequencing centers in the US and worldwide including: Baylor College of Medicine, Washington University, Stanford University, the Whitehead Institute, TIGR, the Sanger Centre (UK), and Généthon (France). The NCBI has a similar list of Genome Centers: **http://www.ncbi.nlm.nih.gov/PMGifs/Genomes/links.html**. Additional genome data may be obtained from these centers.

TIGR was the first to sequence a complete microbial genome (*Haemophilus influenzae*) (5), the first to sequence the complete genome of an Archaebacteria (*Methannococcus jannaschii*) (1), and now plans to be the first to complete the sequence of the human genome. TIGR maintains a very extensive genome sequencing Web site that contains information both from TIGR sequencing projects as well as other genome projects around the world (**http://www.tigr.org/tdb/tdb.html**).

TIGR lists all of the completed microbial genome sequences in GenBank® as well as 54 microbial genomes that are currently being sequenced. TIGR also lists progress on the sequencing of the *Trypanosoma brucei*, *Plasmodium falciparum*, *Arabidopsis thaliana*, and human genomes. TIGR maintains indexes of expressed genes (consensus sequences obtained from ESTs, cDNAs, and transcribed genomic sequences from GenBank) for *A. thaliana*, mouse, rat, rice, zebrafish, and human.

For its own sequencing projects, TIGR provides access to the up-to-date sequence data by file transfer protocol (FTP), similarity search, keyword search, and annotated genome maps. For projects conducted at other sequencing centers, TIGR provides links to sequence data on their Web sites.

EXPRESSED SEQUENCE TAGS

As high-throughput sequencing became routine in the early 1990's, Dr. Craig Venter, then at the NIH, proposed scaling up sequencing using many automated machines and applying it to entire cDNA libraries generated from specific tissue types. Rather than carefully sequencing each cDNA clone, Venter proposed to make single-pass sequences of the 3′ and 5′ ends of many clones, thus creating a collection of ESTs (expressed sequence tags) (1). The NIH initially rejected Venter's proposal, so he founded a private company, TIGR, to pursue large-scale cDNA sequencing. This concept of sequencing ESTs quickly caught on with a number of biotechnology and pharmaceutical companies that developed their own EST sequencing projects (and private databases).

In 1994, Merck (Whitehouse Station, NJ, USA) announced a partnership with the Genome Sequencing Center at Washington University in St. Louis to sequence ESTs from 200 000 cDNA clones and immediately provide the data to the public by submitting it to GenBank. This created yet another level of escalation in the quantity of sequence data flowing into the public databanks—and threatened to render the private EST databanks largely obsolete. As of October, 1999, the EST section of GenBank contains over 3 000 000 sequences, while all of the non-EST GenBank divisions together contain only approximately 700 000 sequences.

The EST databanks now contain almost all of the expressed human genes. This has led to a number of interesting applications of the data. It is possible to discover new members of gene families and human homologs for genes previously identified in other species. Newly sequenced genomic DNA can be compared en masse to the EST databanks by simple similarity algorithms to locate potential gene coding regions. Since many genes are present in multiple copies in cDNA libraries, the relative abundance of these clones in EST libraries from different tissue types provides a rough estimate of relative gene expression levels. The redundancy of the EST databanks has also been used to identify potential sequence polymorphisms that can be used to create single nucleotide polymorphism (SNP) markers (6).

POLYMORPHISMS

Once the human genome has been completely sequenced, informative regions can be re-sequenced much faster and more easily to identify the most common variants of medically important genes in the human population. Diagnostic tools based on polymerase chain reaction (PCR) and other related technologies can then be developed for rapid inexpensive genotype assays as a medical tool. DNA chip technologies are becoming available from companies such as Affymetrix (Santa

Clara, CA, USA) that automate this type of resequencing. Diagnostic sequencing of dozens of disease-related genes will soon become part of routine medical testing.

SNPs are a hot item in current discussions about practical applications of genomics and the Genome Project research. SNPs are simply DNA point mutations—single base pair changes or insertion/deletions—that are present at measurable frequencies in the population. Some SNPs are the most common variations in the human genome, occurring once every 500 to 1000 bp. SNPs exist at defined positions within genomes (sequence tagged sites, STSs) and can be used for gene mapping, defining population structure, and performing functional studies (6). Some SNPs occur within the coding regions of genes—and in fact, some SNPs are the sites of mutations with functional significance (i.e., alleles). SNPs are expected to greatly facilitate large-scale genetic studies concerned with the linkage between sequence variation and heritable phenotypes. SNPs may also become an efficient tool for genetic identification for legal and forensic applications.

SNPs can be used as genetic markers by using a variety of different assay systems including direct sequencing, nucleotide specific PCR primers, ligase chain reaction, DNA chips, or other assay systems that allow rapid identification of alternate alleles in DNA samples. DNA chip microarrays are particularly well suited to SNP assays since large numbers of SNPs can be assayed simultaneously on a single sample. Prototype genotyping chips have already been developed by Wang et al. (12) at the Massachusetts Institute of Technology (MIT) Whitehead Institute (Cambridge, MA, USA), which allow simultaneous genotyping of 500 SNPs. Some advantages of SNPs over other types of genetic markers such as isozymes, restriction fragment-length polymorphisms (RFLPs), variable number of tandem repeats (VNTRs), and simple sequence repeats (SSRs) include: *(i)* unlimited numbers of polymorphic loci; *(ii)* even distribution of loci throughout the genome; *(iii)* markers present within coding regions, introns, and regions that flank genes; *(iv)* simple and unambiguous assay techniques; *(v)* high levels of polymorphism in the population; *(vi)* stable Mendelian inheritance; and *(vii)* low levels of spontaneous mutation. Kwok et al. (6) have observed that approximately 62% of SNPs identified by scrutiny of public sequence databases have heterozygosities exceeding 32%.

DNA CHIPS

DNA chip microarrays are a new technology that is a direct outgrowth of the availability of genome sequence information. A very large number (approximately 100 000) of cDNA sequences or synthetic DNA oligomers are attached onto a glass slide (the chip) in known locations on a grid. An RNA sample is labeled and hybridized to the grid. Then the relative amounts of RNA bound to each square in the grid are measured. DNA chips can be used for the simultaneous monitoring of the levels of expression of all of the genes in a cell or an organism or to study whole genome expression patterns in various tissues during development. It is possible to measure differential gene expression in healthy versus diseased tissue, or the time course of gene expression of a particular cell type upon exposure to a stimulus. The key to this technology is genome-wide sampling—measuring the expression of all genes, not just one or a few.

This type of experiment generates a lot of data—100 000 or more data points for each sample hybridized to a chip. Computational analysis is essential, not just for simple comparisons of expression changes versus background, but for pattern analysis to identify co-regulated (or inversely regulated) genes and to integrate new information about gene expression with the vast databases of existing information about gene function and structure, metabolic pathways, etc. Software for the analysis of DNA chip data is still in its early developmental stages, but it is clear that it will involve large-scale relational databases such as Oracle® (Oracle, Redwood Shores, CA, USA).

COMPARATIVE GENOMICS

Determining the DNA sequence of the complete human genome is merely a starting point; the real goal of the Genome Project is to understand the function of all of the approximately 100 000 human genes. However, it is important to keep in mind that the human is a very poor organism for genetic research because: *(i)* gene knock-outs are impossible; *(ii)* controlled breeding is impossible; and *(iii)* there is not even a comprehensive collection of mutants. Therefore, the complete genomic sequences of many experimental organisms such as mouse, fruit fly, and worm are also needed as experimental tools and as a basis for comparison (comparative genomics). The complete genomic sequences of many prokaryotic organisms (bacteria) have already been determined, and many more are planned or in progress. The collection of complete genome sequences from multiple species and multiple individuals from these species is proceeding rapidly, but the analysis of data at this level has not been adequately addressed by the current generation of computational tools.

The availability of genome sequences from model organisms is creating a new field of comparative genomics, which asks, if a gene has unknown function in humans, then what is the phenotype of a mouse or a fruit fly that has had the corresponding gene deleted? Many functional gene clusters and regulatory interactions are also preserved across species, so experimental results from model organisms can be directly related back to humans.

The sequencing of complete bacterial genomes is now becoming almost a commonplace occurrence. As of October, 1999, GenBank contained the following complete genomes:

Archaebacteria

Aeropyrum pernix
Archaeoglobus fulgidus

Methanobacterium thermoautotrophicum
Methanococcus jannaschii
Pyrococcus abyssi
Pyrococcus horikoshii

Eubacteria

Aquifex aeolicus
Bacillus subtilis
Borrelia burgdorferi
Chlamydia trachomatis
Chlamydia pneumoniae
Escherichia coli
Haemophilus influenzae
Helicobacter pylori
Mycobacterium tuberculosis
Mycoplasma genitalium G37
Mycoplasma pneumoniae M129
Rickettsia prowazekii
Synechocystis sp. PCC6803
Thermotoga maritima
Treponema pallidum

Eukaryotae

Caenorhabditis elegans
Saccharomyces cerevisiae

One of the key objectives of sequencing complete genomes is the ability to make fundamental comparisons between organisms. Virtually all of the important pathogenic microorganisms will be completely sequenced in the next few years. Several new types of analyses have become possible using these whole genomes. One of the most interesting is the ability to state definitively which genes are not present in the genome of a particular organism, and conversely, which genes are present and, therefore, presumably essential in all organisms.

Work on comparative genomics is only just beginning, but there are several interesting examples on the Web. The COG project (Clusters of Orthologous Groups) on the NCBI Web site (**http://www.ncbi.nlm.nih.gov/COG/**) is an exploration of the fundamental similarities between the genomes of diverse organisms (10). Comparison of proteins encoded in seven complete genomes [*E. coli*, *H. influenzae*, *M. genitalium*, *M. pneumoniae*, *Synechocystis* (Cyanobacteria), *M. jannaschii* (Archaebacteria), and *S. cerevisiae* (Eukaryota)] from each of the major phylogenetic lineages allowed the delineation of 722 clusters of orthologous groups. Of these clusters, 95% have known functions; 45% are shared by all three kingdoms (Bacteria, Eukarya, and Archaea), while 29% are shared by just Bacteria and Eukarya; and 17% are shared by just Archaea and Bacteria.

The protein extraction description and analysis tool (PEDANT) at the Munich Information Center for Protein Sequences (MIPS) Web site (**http://pedant.mips.biochem.mpg.de/**) provides comprehensive automated analysis of sets of protein sequences from genome sequencing projects. PEDANT includes all of the predicted proteins from the completely sequenced genomes from GenBank as well as many additional unfinished genomic sequences. PEDANT automatically annotates each protein sequence with the results of similarity searches with FASTA and BLAST, multiple sequence alignments, prediction of secondary structures, transmembrane regions, signal peptides, coiled coils, and the detection of conserved protein domains using PROSITE patterns, BLOCKS, and Hidden Markov Models (HMMs).

Another approach to comparative genomics is being called metabolic reconstruction. Biochemical pathways are delineated, and the genes for each enzyme are found in the genome of an organism. Pathways that appear to be missing enzymes can be investigated in greater detail. It should be possible to determine if an organism truly lacks a biochemical function, or if it performs that function with a gene that is so different in its sequence that it cannot be identified with bioinformatics tools.

The What Is There (WIT) Web site (**http://wit.mcs.anl.gov/WIT2/**) is an approach to metabolic reconstruction created by R. Overbeek, G.D. Pusch, M. Dsouza, N. Maltsev, and E. Selkov of the Argonne National Laboratory and N. Larsen at Michigan State University (9).

The underlying assumption of WIT is that all organisms share (most of) the same metabolic pathways, and they generally perform the same enzymatic functions with proteins encoded by homologous genes. Thus, the process of creating functional annotations for all of the genes in the complete genomic sequence of an organism can be considered as a metabolic reconstruction of that organism. The goal of producing a metabolic reconstruction is to identify which pathways are present in an organism and which genes implement the functional roles. TIGR has made extensive use of WIT in the annotation of its various genome sequencing projects. The creators of WIT hope that members of the scientific community will become curators of metabolic reconstruction models:

> We would like to enable any user, including you, to produce at least a crude overview of which main pathways exist in any organism for which there is a significant amount of genomic DNA sequences available.

GENOME BIOINFORMATICS

The human genome contains about three billion base pairs. GenBank currently

(October, 1999) holds over 3.5 billion base pairs from all organisms. Adding the remaining portion of the human genome is not going to overwhelm the system.

What computational hardware and software would be necessary to perform whole genome comparisons between species to identify all common genes? How about comparing the complete genomic sequences of two individuals of one species to identify genes responsible for various phenotypic differences? Routine whole genome diagnostics was imagined in the "StarTrek" universe of the 24th century, but these type of data may be readily available within one or two decades.

Dr. Eric Lander (director of the MIT Genome Laboratory) has made some predictions about the new kind of biology that will emerge after the human genome is completely sequenced (7). He has compared the human genome sequence to the Periodic Table of chemistry, suggesting that once the genome is sequenced, general rules of information flow in biology can begin to be formulated. The implication is that up until now, biology has been fumbling around without a clear understanding of the fundamental components of organisms. With the new "Periodic Table" of genomic sequences and the associated complete list of expressed proteins, biology will really take off as a rigorous scientific discipline, perhaps sharing some of the mathematical rigor of physics and chemistry.

As biologists, we need to imagine how our work will be different in this Post-Genome Project era. A completely new way of thinking about cloning experiments will be required when every possible gene is available at a computer terminal and various laboratory tools (restriction enzymes, hybridization, PCR, etc.) can be modeled by computer simulations. However, it is important to keep in mind that bioinformatics will not reduce the scope or the need for laboratory research in biology. Quite the opposite, this wealth of new data will lead to an explosion of interesting questions that can only be pursued experimentally. While all of the genes in humans and other organisms will be defined by sequencing projects, most of their functions will only be discovered by experimental work. Biologists will be working far into the future to describe levels of gene and protein interactions within cells and organisms that cannot even be studied by current scientific methods because we lack the complete vocabulary of the genome.

THE GENOME ANNOTATION PROBLEM

The many Genome Projects (human, mouse, *Arabidopsis*, *C. elegans*, microbial, etc.) are churning out genomic DNA sequence at a tremendous rate, far faster than the human staff of the genome sequencing laboratories or the database organizations can hope to analyze and annotate the sequences. As a result, the vast majority of new DNA sequence information pouring into GenBank is annotated automatically. Genes are found in genomic DNA by automatic open reading frame (ORF) detection algorithms and automatically annotated based on similarity searches against the proteins in the databases. However, since the bulk of the proteins now in the GenBank database have had their functional annotations also assigned by these automatic methods, the quality and reliability of this information is increasingly doubtful. For example, it is not very informative to read an annotation that describes a putative protein as similar to another protein of unknown function that has an unsubstantiated similarity to a protein kinase.

These automatic annotation methods have proven to be quite powerful in the case of microbial genomes. Prokaryote protein coding genes can be distinguished with a high degree of accuracy. However, in eukaryote genomes, there are many complexities that cannot be easily overcome by simple-minded algorithms. It requires very careful scrutiny to distinguish between a pseudogene and a member of a multigene family. Transposons and many other types of repeated sequences occur both within expressed protein coding sequences and in noncoding regions.

In addition, the biological knowledge base on which these gene function annotations are based is always changing. If a new function is discovered for a protein, that information must be updated in all of the sequence annotations that have derived their information from that protein. The current structure of GenBank (and other sequence databanks) is not relational, so a change in one record does not cause other related records to be updated.

One way to improve the quality of the annotations of the genes discovered by the genome sequencing projects is for the wider biological community to participate. Biologists could spend a few hours per week studying and annotating genomic sequences relevant to their research. So, if you would like to contribute sequence annotation information, where do you start and what tools will you use? The first step is to obtain some genomic sequence that needs annotation. The various Genome Project Laboratories are now submitting chunks of genomic sequences to GenBank as assembled contigs on a nearly continuous basis. You might pick sequences from an organism, or a chromosome of interest, or look at sequences that contain genes with putative similarity to proteins that you study.

There is clearly a need for experts from a wide variety of biological disciplines to participate in the annotation of the proteins identified by genomic sequencing. Every gene given a more detailed annotation and every mistaken database annotation corrected will have ripple effects throughout the database, since so many gene functions are assigned based on sequence similarity to other genes. The keepers of the public databases will need to develop systems for coordinating this global collaboration on genome annotation.

There are no tools currently available for large-scale participation of biologists in genome sequence annotation. A concept for a genome annotation tool called the Genome Channel is being created by a group known as the Genome Annotation Consortium, which includes members from Oak Ridge National Laboratory, Lawrence Berkeley National Laboratory, Argonne National Laboratory, and collaborators from many of the genome sequencing laboratories. A prototype of the Genome Channel is available on the Oak Ridge Laboratory Web site (**http://compbio.ornl.govtools/channel/**).

The Genome Channel provides a graphical chromosome map interface (karyotype view) to all of the public genome sequencing projects, both completed genomes and those currently underway (Figure 1). Individual contigs from the various sequencing laboratories are shown as color-coded bands at the appropriate positions on the chromosomes. Each contig is directly linked to an information-rich map that includes features such as ORFs and transcription factors, BLAST reports for each predicted gene, the sequencing clones and gel reads that were used to build the contig, and the DNA consensus sequence at that position. Once the prototype of the Genome Channel becomes functional, users from the scientific community will be able add new features or modify the annotations of existing features. Beware, this site makes extensive use of JAVA and may not be compatible with all Web browsers.

THE GENOME PROJECT AND THE INTERNET

So How Does the Genome Project Relate to the Internet?

Clearly the Genome Project involves a huge amount of data that is stored on

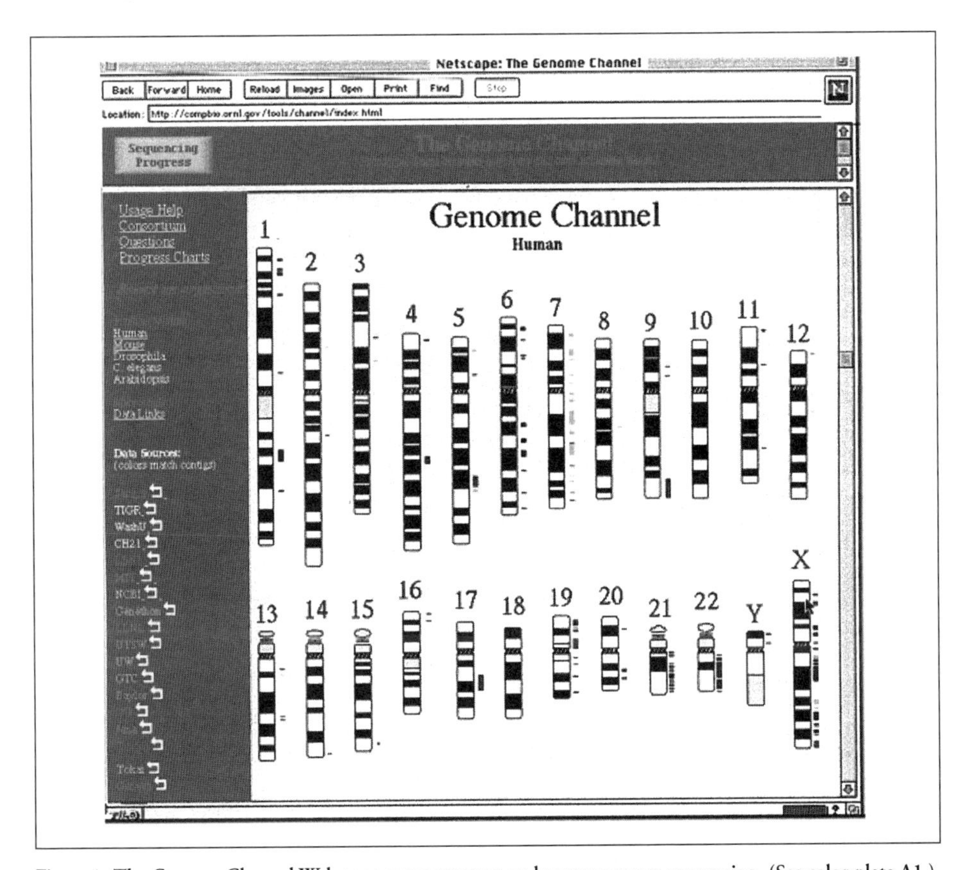

Figure 1. The Genome Channel Web page maps progress on human genome sequencing. (See color plate A1.)

computers all over the world. More than just vast amounts of DNA sequences, the genome project is about developing sets of integrated maps that involve genetic, physical, and sequence data. These maps are linked to information in the public DNA and protein sequence databanks, published scientific literature, images in a variety of formats, and databases that inventory the physical archives of DNA clones, mutant strains of laboratory organisms, hybridization filter grids, PCR primers, etc.

This data can be sorted, annotated, and organized in many different ways using different types of database software, different analysis algorithms, and different forms of interfaces. In many cases, it is this value added processed data that is most useful to the researcher.

- Access to the data is via the Internet, usually via the World Wide Web. New genome-related Web sites are appearing almost daily.
- Genome scientists have an unprecedented level of dependence on the Internet in order to do their work. It is simply impossible to maintain up-to-date copies of all relevant databases within a single laboratory. Larger institutions that do maintain local copies of databases are dependent on high-speed Internet connections to the primary database sites [i.e., GenBank at NCBI, European Molecular Biology Laboratory (EMBL), SWISS-PROT at the University of Geneva, REBASE at New England Biolabs, etc.] to keep their data current.
- The job of annotating complete genome sequences is too large for any single scientific group—or any single approach. The majority of GenBank sequences are now contributed by the large-scale genome project laboratories (and that fraction will be increasing), which of necessity use automated annotation software. In order to achieve the best possible database annotation, the job needs to be distributed among the entire worldwide scientific community so that all investigators can contribute their own insights to their genes of interest. This worldwide collaborative project relies heavily on the Web.
- In many cases, genome project data centers are leading the world in developing new Internet tools for accessing data.

One of the key ways that the Genome Project is bringing new types of data to the Internet is in the area of genetic maps. The Genome Project was conceived as a series of integrated maps at ever higher resolutions—first, genetic maps that included both mutations (with observable phenotypes) and molecular markers, then, physical maps at the level of chromosome bands, then 100 megabases yeast artificial clones (YAC) clones, 20 kb cosmids, and finally, the DNA sequence itself.

Some examples of databases that take advantage of the graphical and interactive features of the Web are the chromosome map-based Genome Channel at the Oak Ridge National Laboratory (**http://compbio.ornl.gov/tools/channel/index.html**) and the human genome transcript map at NCBI (**http://www.ncbi.nlm.nih.gov/SCIENCE96/**) (Figure 2).

The *B. subtilis* genome sequencing project has a Web site at the Institut Pasteur (**http://www.pasteur.fr/Bio/SubtiList.html**), which provides a clickable map of the *B. subtilis* chromosome that leads to an integrated physical and genetic map of the genome (Figure 3). The map is fully annotated with every ORF, which are in turn linked to information about each gene, including functional information about the protein product, its molecular weight and isoelectric point (pI), and links to other databases such as SWISS-PROT, PROSITE, etc.

GENOMICS AND PHARMACEUTICAL RESEARCH

The essence of pharmaceutical research has always been identifying a disease or symptom and finding drugs that cure or improve this condition. The discovery of disease-causing microorganisms at the dawn of the era of modern medicine created a wave of pharmaceutical innovation as researchers sought compounds

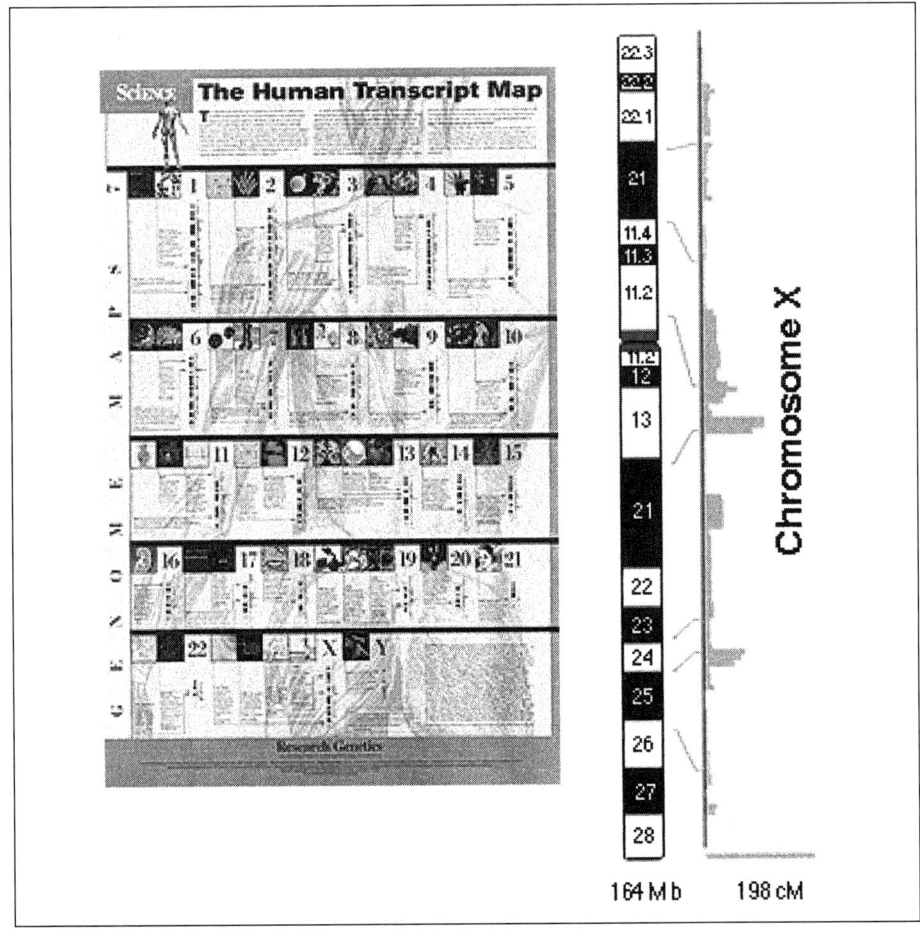

Figure 2. NCBI Web page for the human genome transcript map and a sample map of the X chromosome. (See color plate A2.)

that attacked these pathogens. The Genome Project is creating a similar revolution in the way scientists think about drug research and discovery. Now it is increasingly possible to identify particular genes responsible for (or involved in) the many diseases that cannot be treated with antibiotics.

Information about genes and the molecules they interact with can be used to rationally identify, design, create, and test new drugs. Every new gene sequence that can be associated with a specific function, whether it is a disease phenotype or a metabolic pathway or a regulatory system, becomes a potential target for the design of new drugs. Then, rather than randomly screening potential drug chemicals for effects on a disease, drugs can be intentionally designed to interact with a gene target. Drugs can act at the regulation of gene expression at the DNA level, affect its mRNA product (ribozymes and antisense technologies), or interact with the gene's protein product. Since the DNA sequence reveals the amino acid sequence of the protein, specific antibodies, competitors, or inhibitors can be designed to interact with the protein (to either reduce or enhance its function), or the protein can be replaced by gene therapy if it is damaged or deficient. Alternately, a variant form of a protein may be engineered as a therapeutic agent. The process of designing and screening compounds that interact with a target gene or protein is known as combinatorial chemistry and high-throughput screening. These are areas of chemistry and experimental cell biology that also can be considered part of bioinformatics, but exceed the scope of this book.

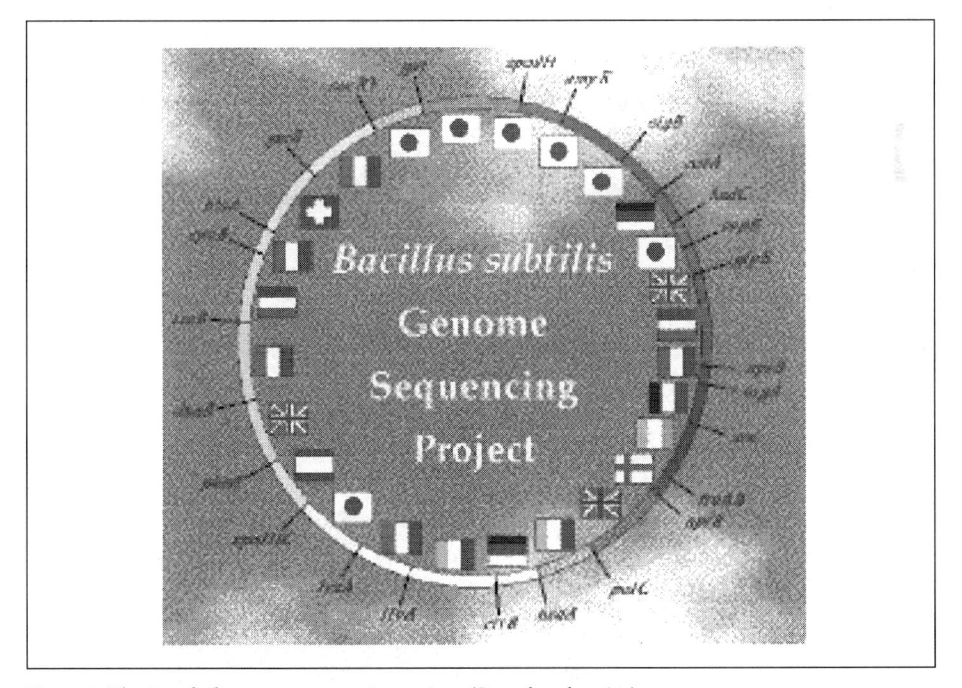

Figure 3. The *B. subtilis* genome sequencing project. (See color plate A3.)

Pharmaceutical research companies are hungry for new genome data and new bioinformatics tools that can lead to the discovery of new genes. This is a highly competitive field, since the production of effective drugs for diseases such as cancer, heart disease, or obesity would be fantastically profitable. Genomes are rapidly being sequenced and genes identified. The most challenging bioinformatics problems in pharmaceutical research are identical to those faced by academic researchers: predicting protein structure from amino acid sequence, identifying regulatory interactions between genes, and identifying distant homologues both across species and across families of related proteins.

SNPs have been proposed as an ideal tool for the emerging discipline of pharmacogenomics. This new discipline hopes to use genetic information to fine-tune patient diagnosis and treatment. Large-scale analysis of the associations between the effects of drugs and genetic markers may allow physicians to match drugs to the genetic makeup of individual patients to better predict beneficial and harmful effects. It may also improve the speed and effectiveness of clinical trials by prescreening patients to identify those who are likely to respond well to new drugs. Linkage analysis with large numbers of SNPs could identify genes and specific combinations of alleles responsible for many complex polygenic diseases. It may also permit the identification of subtypes of complex syndromes such as hypertension, which are known to have different causes in different people.

REFERENCES

1. **Adams, M.D., J.M. Kelley, J.D. Gocayne, M. Dubnick, M.H. Polymeropoulos, H. Xiao, C.R. Merril, A. Wu, B. Olde, R.F. Moreno, et al.** 1991. Complementary DNA sequencing: expressed sequence tags and human genome project. Science *252*:1651-1656.
2. **Bult, C.J., O. White, G.J. Olsen, L. Zhou, R.D. Fleischman, G.G. Sulton, J.A. Blake, L.M. FitzGerald, R.A. Clayton, J.D. Gocayne, et al.** 1996. Complete genome sequence of the methanogenic archaeon, *Methanococcus jannaschii*. Science *273*:1058-1073.
3. **Collins, F.S. and D. Galas.** 1993. A new five-year plan for the U.S. Human Genome Project. Science *262*:43.
4. **Collins, F.S., A. Patrinos, E. Jordan, A. Chakravarti, R. Gesteland, and L. Walters.** 1998. New goals for the U.S. Human Genome Project: 1998-2003. Science *282*:682-689.
5. **Fleischmann, R.D., M.D. Adams, O. White, R.A. Clayton, E.F. Kirkness, A.R. Kerlavage, C.J. Bult, J.F. Tomb, B.A. Dougherty, J.M. Merrick, et al.** 1995. Whole-genome random sequencing and assembly of *Haemophilus influenzae* Rd. Science *269*:496.
6. **Kwok, P.Y., Q. Deng, H. Zakeri, S.L. Taylor, and D.A. Nickerson.** 1996. Increasing the information content of STS-based genome maps: identifying polymorphisms in mapped STSs. Genomics *31*:123-126.
7. **Lander, E.S.** 1996. The new genomics: global views of biology. Science *274*:536-539.
8. **NIH: Understanding Our Genetic Inheritance.** 1990. The U.S. Human Genome Project, Fiscal Years 1991-1995. DOE/ER-0452P, NIH Publication No. 90-1590.
9. **Overbeek, R., N. Larsen, G.D. Pusch, M. D'Souza, E. Selkov Jr., N. Kyrpides, M. Fonstein, N. Maltsev, and E. Selkov.** 2000. WIT: Integrated system for high-throughput genome sequence analysis and metabolic reconstruction. Nucleic Acids Res. *28*:123-125.
10. **Tatusov, R.L., E.V. Koonin, and D.J. Lipman.** 1997. A genomic perspective of protein families. Science *278*:631-637.
11. **U.S. DOE Human Genome Program.** 1999. HGP leaders confirm accelerated timetable for draft sequence. Human Genome News, 10, No. 3-4.
12. **Wang, D.G., J.B. Fan, C.J. Siao, A. Berno, P. Young, R. Sapolsky, G. Ghandour, N. Perkins, E. Winchester, J. Spencer, et al.** 1998. Large-scale identification, mapping and genotyping of single-nucleotide polymorphisms in the human genome. Science *280*:1077-1082.

3 The Role of Computers in the Biology Laboratory

COMPUTERS IN THE LABORATORY

The availability of computers in the laboratory is changing the way biologists work—in both profound and trivial ways. Computers exist in the laboratory in two different forms: as components hidden within (or attached to) other machines, and as multipurpose workstations.

Computers are now found with greater frequency in laboratories as integral parts of automated machinery. One aspect of this trend is the computerization of familiar machines. This trend can be seen at home (cars, microwave ovens, and VCRs) as well as in the laboratory (centrifuges, polymerase chain reaction [PCR] machines, etc.). The computers allow these devices to remember a variety of programs that involve executing complex sets of instructions. There are other machines, such as pipetting robots for the automated assembly of reactions, which could not exist at all without computer technology. Computers have made the most significant impact on laboratory work in the area of automated data collection. All manner of analytical equipment can now be set up and left alone while computers run preprogrammed analytical tests, collect data, and process that data to produce results that can be directly used by the scientist. Once again, there are some devices that have been in the laboratory for many years that have now been fitted out with computers for data collection (high-performance liquid chromatography [HPLC], scintillation counter, etc.), while others, such as the automated DNA sequencer, are true products of the computer revolution.

The overall effect of the automation of the laboratory is the increase in the speed at which data is generated. Disciplines such as DNA sequencing, which are limited primarily by data collection, have improved their productivity tremendously. Another trend that is having a major impact on the use of computers in biology is the scaling up of the research process. In corporate, university, and government laboratories, research groups are growing larger, and technical functions are being spun off to core laboratories or to private subcontractors—so that those

Bioinformatics
By Stuart M. Brown

who deal with automated machinery need to process increasingly large amounts of data for a variety of different experiments. Similarly, scientists who once spent a lot of time on repetitive laboratory tasks, now spend more time analyzing data and planning experiments. This type of high-throughput generation of information in the laboratory is putting increasing demands on bioinformatics.

Biology as a whole has been sped up by the more rapid availability of analytical data—yet scientific discovery is still limited by the ability of researchers to set up complex experiments and to analyze the streams of data as they flow in. Analysis and understanding are still the province of the human mind, but the computer is beginning to play a role in this more abstract area as well.

Computer programs have been available for several decades that can analyze groups of numbers. Spreadsheets and statistical programs are useful to scientists who work with quantitative information—in fact, the same tools are used by accountants and business managers in all types of organizations. The real advance of modern computers over these simple number-crunching applications has come in the visualization of data. Much biology data is inherently visual rather than quantitative in nature, from genetic maps to 2-D protein gels. The scientist can comprehend biological complexity better as images than as arrays of numbers. As the power of personal computer workstations have increased, so has their ability to render data as complex graphical output—from multidimensional graphs to real-time simulations of cellular processes. Better visualization of data and of conceptual models can lead to better science, either through more rapid comprehension of data or by providing a perspective that allows for the development of entirely new insights.

The desktop computer has become a ubiquitous part of every molecular biology laboratory. Laboratory computers are used for word processing, storing of experimental records, and sequence analysis. For example, computer-generated maps and simulations of restriction digests are now an essential part of the planning for every step in a cloning project. By keeping laboratory notes on the computer, data can be shared much more easily among members of a laboratory group, and the documentation and publication of results are greatly facilitated. A new concept in the use of computers in the laboratory is the electronic laboratory notebook. A pioneer in this area is the Gene Inspector™ program from Textco (West Lebanon, NH, USA). It remains to be seen if many laboratory scientists will be willing to abandon their trusty laboratory notebooks filled with careful (or illegible) notes, taped-in Polaroid photos, and smudged scribbled-on photocopies of protocols in favor of a truly electronic notebook.

Another computing innovation, which is trickling down from large-scale genome project laboratories to individual investigators, is the creation of laboratory databases for sequence, clone, and sample information. Many molecular biology laboratories have -80°C freezers filled with thousands of tubes of frozen *Escherichia coli* cells that contain various plasmid constructs. If the laboratory is well organized, then there is some map or notebook that describes these samples.

It would be nice, however, to have a computer database that describes each plasmid, and provides its sequence, restriction map, and cloning history. The program Vector NTI™ from InforMax (Bethesda, MD, USA) is designed to provide this laboratory database function. It also can be used to create a shared database across a multilaboratory work group, a department, or an entire institution. Genetics Computer Group (GCG) is also trying to move into this area with a new product called SeqStore™, which includes a custom Oracle® database (Oracle, Redwood Shores, CA, USA) and a Web-based interface.

The most revolutionary new development for computers in the laboratory is access to the Internet. Now, ordering reagents and equipment, looking up bibliographic references, or consulting genetic maps can all be accomplished in a few minutes from the laboratory computer. The Web sites from several manufacturers of biological products provide rapid searches for reagents offered by a broad range of different vendors, immediate supplier information, and quick direct price quotes. Ordering laboratory supplies over the Web can replace shelves of printed catalogs that are limited by the nature of their medium, expensive to produce, and out-of-date even before they are printed.

All of the major biological databases (including GenBank®, European Molecular Biology Laboratory [EMBL], SWISS-PROT, Protein Information Resource [PIR], Genome Database [GDB], MEDLINE, and all of the ACDB species databases managed by the United States Department of Agriculture [USDA]) have made their information available on the Web. Simple direct access to these databases over the Web from the laboratory has changed the way biologists work. Access to unlimited information is empowering. It doesn't just speed up scientific work, it creates new opportunities—avenues for inquiry that would not otherwise be pursued. Now it is routine to immediately check every fragment of DNA sequence as it is determined in the laboratory for similarity to anything in the GenBank database. With the complete genomic sequences of yeast, *Caenorhabditis elegans*, *Drosphila*, the many prokaryotes now available, plus the inclusion of ever larger collections of expressed sequence tags (ESTs) collected from humans and a variety of other species, it is increasingly unusual to discover a new sequence that lacks any homologs in the database. This sequence information is used to shortcut the traditional process of cloning genes. It is now common to clone a gene by using sequence information from related genes to build PCR primers, or to use computer databases to identify existing cDNA or genomic clones that contain the gene of interest, and bypass the entire process of building and screening libraries.

The trend toward more network-oriented computing environments is likely to accelerate in the near future. As more and more information resources and sequence analysis tools become available over the Internet, the importance of the applications installed on a particular computer declines. In fact, the processor speed and hard drive capacity of laboratory computers is rapidly becoming irrelevant, since they now function primarily as hosts for Web browser, e-mail, and terminal emulation clients.

Predicting the future of the Internet is a good way to destroy one's credibility, but some trends are so powerful that they cannot be ignored by biologists who expect to still be working in 5 to 10 years. First, more and more information that was once only available in print (newspapers, journals, and textbooks) will be placed on the Web. Second, many applications that traditionally were installed and run on a single desktop computer or a local mainframe machine will be available on the Internet. Third, users will be charged for access to some information and applications. Fourth, information that once lived only on dedicated computer systems running obscure database programs (such as your medical, insurance, credit card records, or the airline reservation system) will be accessible over the Internet. Access to this information will be controlled by passwords and/or authentication of users with ID cards, fingerprint readers, or other security schemes. The result will be a sea of information that is available to those with the access codes, regardless of their location on the planet.

A BRIEF HISTORY OF THE INTERNET

It may seem like too much of a reach to predict major changes in the culture of scientific research due to advances in Internet technology until one looks back on the ever accelerating pace of change that is the history of the Internet.

The Internet was created by scientists, and scientists continue to lead the way developing new Internet technologies and finding new ways to use the Internet in their work. Molecular biologists are particularly dependent on the Internet since the DNA and protein databases are so huge and updated so frequently that access

"Hello, is this the Internet?"

via the Internet is the only practical way to use this data. The Internet has grown so quickly over the past few years that at any given time, more than half of the users are new. Scientists have been leading the pack, but it is still useful to provide a little historical perspective. It is important to realize just how rapidly the Internet has evolved, and how short a time it has taken for us to take it for granted.

The concept of the Internet originated in a study in 1962 by Paul Baran, of the RAND Corporation, that was commissioned by the U.S. Air Force on how to maintain command and control over its missiles and bombers after a nuclear attack (1). Baran proposed a computer-to-computer communications network based on "packet switching." The network would function by breaking down data into small packets that were labeled to indicate the origin and the destination of the information and the forwarding of these packets from one computer to another until the information arrived at its final destination. If a computer or a communications line in the network failed, the packets could be sent on a different path to their destination. If any packets were lost in transmission, the message could be resent by the originator.

In 1968, The Advanced Research Projects Agency (ARPA), a unit of the U.S. Department of Defense contracted Bolt, Beranek, and Newman (BBN) to build a prototype packet-switching network. The first network, known as ARPANET, was constructed in 1969, linking four nodes: the University of California at Los Angeles, the Stanford Research Institute, the University of California at Santa Barbara, and the University of Utah.

The first e-mail program was created in 1972 by Ray Tomlinson of BBN.

In 1974, Vinton Cerf of Stanford and Bob Kahn of DARPA published a paper describing the TCP/IP (Transmission Control Protocol/Internet Protocol) protocol, which would allow computers on diverse networks to interconnect and communicate with each other. Cerf and Kahn (2) coined the term "Internet" to describe this protocol and the potential for it to be used to create an interconnected network of networks.

Dr. Robert M. Metcalfe, working at Xerox PARC (Palo Alto Research Center), developed a communications standard known as "ethernet" in 1973 (4), which allowed coaxial cable to move data extremely fast within a Local-Area Network (LAN).

USENET (the worldwide news group bulletin board system) was created by Steve Bellovin, a graduate student at University of North Carolina, and programmers Tom Truscott and Jim Ellis in 1979.

The University of Wisconsin created the Domain Name System (DNS) in 1983. This allowed information to be directed to a particular computer based on a domain name (the now familiar **www.company.com**), which would be translated by the server database into the corresponding IP number. This made it much easier for people to access other servers because they no longer had to remember IP numbers.

In 1986, the National Science Foundation launched NSFNet, which was a new network managed by the nonprofit Merit Corporation, separate from ARPANET, but connected to it. NSFNet was a high-speed transcontinental network that connected five supercomputing centers through a T1 line backbone (1.5 Mbps). In turn, the centers made their facilities available to universities in their region, effectively making the network completely decentralized.

By 1990, the NFSNet backbone was upgraded to T3 lines (45 Mbps). ARPANET was disbanded, and it was fully replaced by the NSFNet backbone.

In 1991, the University of Minnesota developed an easy way to organize text information on the Internet called Gopher. Gopher was soon used to provide access to all kinds of information stored on servers in universities, libraries, nonclassified government sites, etc.

Also in 1991, Tim Berners-Lee at CERN (European Laboratory for Particle Physics) in Geneva implemented a hypertext system known as the World Wide Web (WWW) to provide efficient information access to the members of the international high-energy physics community. In 1993, Marc Andreessen led a team at NCSA (The National Center for Supercomputing Applications) at the University of Illinois which developed a graphical user interface to the WWW called Mosaic.

From 1990 to 1995, a number of other private TCP/IP network backbones were built as government and private organizations and required faster and more reliable network service than was available on NFSNet. These included UUNET, PSINET, Sprint, and MCI. The NSFNet backbone handled less and less traffic and was finally shut down on April 30, 1995, leaving the commercial backbones to do the work that it started.

Also in 1995, the commercial dial-up online services, Compuserve, America Online, and Prodigy begin to provide access to the Internet for their customers. Netscape became a public corporation and began to distribute its Netscape Navigator Web browser to the public. The Netscape IPO on August 9 was the 3rd largest ever NASDAQ IPO share value. Beginning 14 September, a $50 annual fee was imposed by INTERNIC for the registration of .COM, .ORG, and .NET Domain Names.

It is very difficult to fully quantify the growth of the Internet as a social phenomenon over the time period from 1995 to the present, but it has clearly changed from a data network serving primarily scientific and technical people to a means of mass communication and global economic activity. There are some statistics that indicate the exponential growth of the Internet—which is even more impressive than the growth of sequence information in GenBank. The number of web sites has increased from approximately 100 000 in January of 1996 to over 1 million in January 2000 (see Table 1). Over this time period, internet traffic has increased at a fairly steady rate of 30% per *month*, i.e., network traffic is doubling every 100 days. There were approximately 10 million hosts (computers with registered IP addresses) on the Internet in January of 1995 and 72 million in January, 2000 (Internet Software Consortium: www.isc.org). Nua Internet Ltd. (Dublin, Ireland) estimates that a worldwide total of 300 million people are on

Table 1. Some Data on the Exponential Growth of the WWW

Date	Number of Web Sites
12/93	623
12/94	10 022
12/95	100 000
12/96	603 367
12/97	1 681 868
12/98	3 689 227
8/99	7 078 194

Data from Hobbes' Internet Timeline, Robert H. Zakon

the internet as of March, 2000, versus approximately 20 to 25 million in 1995. This may reflect the growing number of people who now access the Internet from schools, libraries, and other public computers, rather than from their own dedicated machines.

USING COMPUTERS FOR BIOLOGY RESEARCH

Working with computers is frustrating! This is a universal truth that has not changed a bit in the years that I have spent working, playing, and cursing at computers. It is not likely to change in the next few decades. They are simple, mindless machines that will not do what you want them to unless you provide exact instructions in the correct order, using precisely the spelling, punctuation, letter capitalization, etc. that they are programmed to accept. Computer hardware gets faster, cheaper, and more technologically sophisticated every year, but they do not ever seem to get easier to use. Graphical User Interfaces (GUIs) such as the Macintosh® and Microsoft® Windows® have simplified the routine use of personal computers, but this only hides the complexity behind a friendly mask. The ownership and operation of a personal computer is still a much more time-consuming hobby than a television or a typewriter. Even an automobile, with it tangled variety of electronic, hydraulic, and mechanical subsystems, is still much easier to use and maintain than a computer. Molecular biology computing frequently requires the personal computer user to look behind the GUI mask and deal with the ugly reality of obscure configuration files, incompatible data formats, etc. The biologist performing complex sequence analyses often must work with mainframe machines that use some dialect of the text-based UNIX® or VMS® operating systems.

The most important single piece of advice that I offer to my students is this: "Be persistent!" Do not expect to zip through any computational task in a few minutes. Whatever analysis you are trying to do is possible if you carefully read the relevant help files, try different variations on your commands and settings, explore all of the options in the program that you are using, etc. The key to becoming a skillful computer-using biologist is not memorizing the commands that you need to perform a given task, but rather learning the general skills of getting help on different computer systems and finding ways to work around roadblocks. Scientists move from laboratory to laboratory and university to university, and computer systems change, Web sites go on-line and are discontinued.

Use frustrating experiences with computers as a chance to learn general skills about how to solve problems with computers rather than just focusing on the need to get the sequence analysis done and the results printed for that grant proposal deadline tomorrow.

The computer-savvy scientist needs to use computing tools for a variety of research tasks, from assembling gel reads or automated sequencer runs for a sequencing project, to the design of PCR primers or to the creation of molecular phylogenies of genes from related species. The secret is to know the best tools for the job, but also to be able to use the tools at hand (or to know how to get the tools you need). Any truly computer-savvy scientist should be able to sit down at an unfamiliar network-connected computer or terminal anywhere in the world and (with an appropriate amount of trial and error) be able to read and send e-mail from their home account, conduct a literature search, find a DNA sequence in a databank, design a PCR primer, etc.

TYPES OF COMPUTERS

It is important to define some terms that are used rather loosely in discussions of bioinformatics. The majority of this book is devoted to discussions of various computing solutions to molecular biology problems. The term **algorithm** is used often to imply an unambiguously specified method for solving a problem. In this context, an algorithm may be thought of as a computer program, but it is more accurately described as the general strategy that was used by the programmer in writing the program. There may be a variety of different programs based on a single algorithm, or a computer program may offer the choice of a variety of algorithms to solve one type of problem. However, unlike most discussions of bioinformatics algorithms, this book will avoid using the precise language of mathematics and computer science. Instead, algorithms will be explained in plain English using liberal amounts of metaphor and analogy. The goal is to give the biologist a general idea of how a given computer program approaches a problem, rather than a precise understanding of its mechanics.

A **personal computer** (PC) or **desktop computer** is a Macintosh or an IBM-compatible PC running some version of the DOS®/Windows operating system. The PC is designed as a single-user machine with a single processor. PCs with Intel processors and Windows operating systems have come to dominate the personal computer market for both home and business machines, but the molecular biology community has long favored Macintosh computers. Due to a series of management, marketing, and public relations disasters at Apple Computer, this balance began to shift in 1997. However, in 1998 and 1999, Apple regained significant market share in both the home and academic market. It remains to be seen if scientists and scientific software developers will continue to support the Macintosh.

As a practical matter, both Macintosh and Windows computers can function equally well as Web browsers or to access mainframe computers via telnet or X-

Windows interfaces. They both have excellent word processing, spreadsheet, database, and graphics programs. A skilled user can certainly get the job done on either type of machine. For the beginner, Macs are generally easier to set up and learn to use, but not if you are denied technical support from your institution. Overall, it is probably best to take the path of least resistance and use the type of computer most common in your department or institution.

A **mainframe** computer is a multiuser system generally running some version of the UNIX or VMS operating system. These machines are also called minicomputers in the lingo of true computer scientists. Biologists generally do not make use of real mainframes running operating systems such as IBM's parallel processing OS/390™ or its more traditional MVS/ESA and VM/XA. A mainframe may have a single processor or multiple processors. It may be accessed from dedicated dumb terminals (a screen and keyboard without any processor or internal memory), from X-terminals, or from personal computers by the ethernet over a LAN or over the Internet. It supports TCP/IP, X-Windows, and possibly other communications protocols.

There are several other types of computers that fall into an intermediate classification between the personal computer and the mainframe. The vague general term **workstation** is often used for these intermediate types of machines. Workstations include small mainframe machines dedicated to a single user such as Sun SPARC, DEC Alpha, and Silicon Graphics machines running various versions of UNIX. High-powered PCs running Windows NT® or LINUX can also be considered workstations. Also, Macintosh machines running A/UX (a version of UNIX) or even multiprocessor Macs running the MacOS might be called workstations.

NETWORK COMPUTERS

At universities and corporations, researchers rely on desktop computers (Macintosh and IBM-compatible PCs) that are connected via cables into a large network. Also on such networks are servers, powerful multiuser computers that provide campus-wide services such as e-mail, databases, domain name service (used by WWW browsers), USENET News, etc. Until recently, there was a clear distinction between applications that ran entirely on a desktop machine (word processing, spreadsheets, drawing, and presentation programs, etc.) and applications that relied on the network (e-mail, WWW browsers, USENET News readers, etc.). However, many traditional stand-alone programs have added network functions such as a calendar program that compares your personal schedule over the network to a master schedule for your research group, or a database/spreadsheet program that allows other users to remotely access (or even modify) your data. In the molecular biology area, shared databases of plasmids and freezer stocks are gaining popularity.

The next step in this process is likely to be the advent of true network computers (NCs) that will have no (or minimal) internal hard disk storage, limited processor power, but optimized Web browsing functionality. These NCs would connect to a server that would provide a browser–operating system and a wide range of applications to each user. NCs are particularly well suited for the research laboratory, where each machine is used by many different people, and system maintenance is a dreaded chore. The server could provide each user with a secure file storage area and extensive user preferences. A researcher could then have password-protected access to their personal files, favorite applications, and desktop configurations (perhaps even a choice of operating systems) from any NC in the institution and perhaps from any computer in the world with an Internet connection or a modem.

Apple Computer introduced a type of network computing system called NetBoot in mid-1999 that uses Mac OS-X Server and iMacs as the NCs. Some university computing laboratories have adopted this system, but it has not been popular among laboratory scientists.

There are many advantages to an NC system for both the individual user and the institution. A huge amount of time can be saved that is now spent on the maintenance of a fleet of PCs, each with different hardware and software configurations. NCs will be cheaper than high-powered PCs and will not need to be frequently upgraded. An additional benefit would be substantial savings in software fees, since the number of copies of each application need only be equal to the peak usage rate for that application, rather than the actual number of computers on the network. Furthermore, all users of the system would always have access to the same version of each software program, so incompatible file formats will no longer be an issue. A centralized backup system would prevent the loss of data through hardware crashes or accidental deletion of files from shared machines. It might also be possible to share the combined computing power of many servers among users, so that everyone will have access to as much power as they need for any given computation without having to invest in their own high-powered machine.

Universities and other research institutions are likely to be early adopters of this network computing technology due to the cost savings in hardware, software, and technical support. This goes hand-in-hand with a new generation of molecular biology computing programs that rely on the network to deliver mainframe computer power to desktop machines.

For the home computer user, it is not likely that true NCs will become a popular alternative to the current crop of inexpensive, high-powered machines. However, the Internet is assuming an ever larger role in many common computing tasks. Essentially, any program can be moved from an individual computer onto the Internet. For example, the popular TurboTax/Macintax tax programs from Intuit were available on the Web in the 1998 tax season. Sun Microsystems has announced that it will offer a complete word processing and office productivity package called StarOffice™ over the Web in 2000. Even Microsoft is considering the possibility of offering a Web-based version of its ubiquitous MS Office™ suite of applications (Word/Excel/PowerPoint). Many computer games now include the ability to play over the Internet against (or in teamwork with) other computer users. There is a rapidly expanding market for services offering Web-based e-mail, file storage, or automated backup of data from PCs to Internet servers. This service potentially allows people to access their personal data files from any Internet-connected computer. Many popular applications are now using HyperText Markup Language (HTML) as a standard file format both for user-created documents and for-help files. This allows the nearly seamless integration of information stored on the computer's hard drive with information on the Web.

WORKING WITH MAINFRAMES, PERSONAL COMPUTERS, AND THE WEB

One of the most frustrating things about the current state of computer sequence analysis is the diverse collection of tools that must be used in the course of a routine project. Some aspects of sequence analysis are best accomplished using a program such as that offered by GCG, which is run on a mainframe computer, accessed from a desktop PC with a terminal emulator program such as NCSA Telnet. Other aspects of the analysis can only (or best) be accomplished over the Internet using a Web browser such as Netscape Navigator. Still other parts of the project, such as the annotation of formatted alignments or the printing of circular plasmid maps, require (or are much easier) using stand-alone programs on a desktop computer.

The scientist must jump from one computing platform to another, copying and pasting the relevant data, dealing with inconsistent file formats, different operating system commands, different keyboard mappings, etc. The complexity of the process of using computer sequence analysis tools is a significant barrier to correct analysis. In other situations, there is a significant overlap between the molecular biology functions currently available on the Web, in mainframe computing packages, and as programs that can run on desktop computers.

For example, a Basic Local Alignment Search Tool (BLAST) search can be performed in at least seven different ways: *(i)* on the National Center for Biotechnology Information (NCBI) Web server; *(ii)* by e-mail to NCBI; *(iii)* using GCG on a mainframe computer with a local database (with text, X-Windows, or Web interface); *(iv)* using GCG to send the job to NCBI's computer (NETBLAST); *(v)* using a stand-alone version of BLAST on a mainframe; *(vi)* using a stand-alone version of BLAST on a PC; or *(vii)* using MacVector (or another desktop computer program) to send the job to NCBI's computer. When the same function is available from a variety of computer platforms, what is the best choice?

In this case, the best choice depends on what resources are available, how many searches are going to be needed, and what will be done with the search results. For a single search, where no additional analysis of the results is planned, then the NCBI BLAST Web server is probably the best solution. If graphical Internet access is not available, then the BLAST e-mail server may be the only option. For confidential data, transmission over the Internet to NCBI may be a security risk, so a local mainframe system with a local database may be the only solution. If the results of a BLAST search will become the first step in a more complex analysis, such as the creation of a multiple alignment, then the use of an integrated package such as GCG or MacVector presents a distinct advantage.

Some general guidelines are:

1. Use the Web for quick, stand-alone analysis or when it offers a unique resource.
2. Use a mainframe for large, multistep analysis projects and the analysis of large data sets. Mainframe programs such as offered by GCG are the only option for the multiple alignment of large numbers of long sequences.
3. Use stand-alone PC (or Macintosh) programs when high-speed computational power is not required and greater ease-of-use is desired. Also, desktop computer programs usually provide better graphics and more flexibility to reformat and polish output for final publication.

As a general workflow outline, bring initial results from Web-based tools into a mainframe package for integrated analysis, then move the final results to a desktop program for more attractive formatting.

We are currently in the midst of a major shift in bioinformatics software. Previously, investigators who needed computation-intensive analyses of DNA sequences were forced to use programs such as GCG on a mainframe computer and to learn the arcane command line language necessary to control those programs. Those investigators who had more modest computational requirements, but preferred a graphical, user-friendly interface, generally chose a rather expensive desktop program such as GeneWorks, MacVector, or DNASIS (Hitachi Genetic Systems, Alameda, CA, USA). Web-based graphical interfaces are now becoming available for GCG and other mainframe programs. The Web browser can then provide a universal interface that can include rich graphics without demanding a lot of processing power from the desktop machine. The Web is emerging as a universal computing tool to access databases and mainframe computer programs. This will replace text-only access to programs like GCG and spur the growth of a new generation of sequence analysis software specifically designed to provide an easy-to-use interface that will link databases with sequence analysis programs.

WHY GCG?

Why is GCG mentioned so frequently in this book? Very simply, GCG has become the industry standard for bioinformatics (Figure 1).

GCG, also know as the Wisconsin Sequence Analysis Package (7), is the software product of the Genetics Computer Group, a private company located in Madison, Wisconsin (now a division of Oxford Molecular Group PLC, Oxford, UK). GCG is the outgrowth of a molecular biology computational package that was initially developed by scientists (and computer programmers) at the University of Wisconsin (3). GCG has become the dominant product in use today for molecular biology computing.

GCG is popular with scientists for four reasons: *(i)* it contains the most comprehensive set of sequence analysis tools; *(ii)* the programs are kept up-to-date and follow the current consensus of opinion among computational biology theorists; *(iii)* it is designed to run on very high-powered mainframe computer systems; and *(iv)* a site license is inexpensive for large institutions. This low cost (approximately $5000 for one mainframe with an unlimited number of users) is especially important when compared to the price of molecular biology programs for the PC ($1000–$2000 per computer), which have fewer functions and can be very slow for computationally intensive operations like database searches.

GCG is designed to run on large computers (servers) that use the UNIX operating system. Individual users of the software then connect to the server via telnet (or X-Windows) from terminals or PCs. In this way, a single software package can serve an entire university. This is necessary because of the immense size of modern molecular biology databases. The GenBank database of DNA sequences is currently (as of October 1999) over 18 gigabytes, which represents more than 4

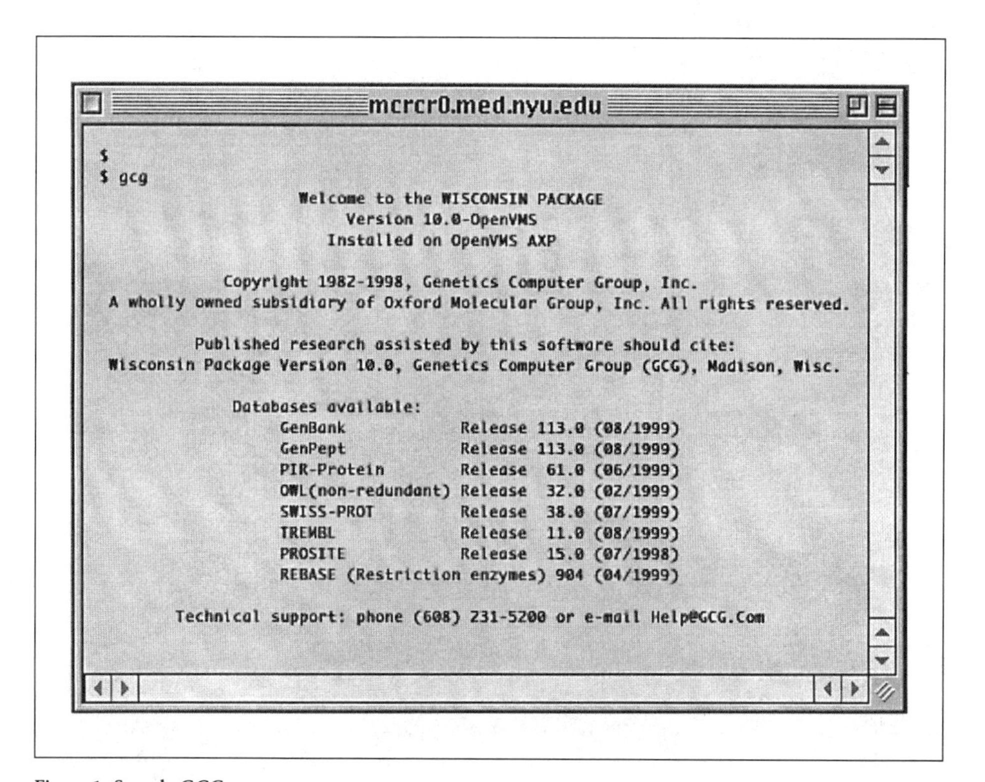

Figure 1. Sample GCG screen.

billion bases. Only a very powerful computer can search this huge database in a reasonable amount of time, and the disk storage requirements for the databases can be met by spreading the costs across a large community of users. However, since at any given moment only a few users are performing processor-intensive computations, every user has a tremendously powerful computer at their disposal.

GCG is actually not a single program but a suite of over 100 programs that all share a similar look and feel. Each program performs a specific function, and the programs are designed to work together to streamline analysis through a variety of steps. GCG initially relied on a command-line-interface (similar to the DOS operating system for IBM-compatible PCs), which has a steep learning curve, but is ultimately very powerful and efficient for experienced users. This interface requires the user to know what program is needed, what options are available for that program, and what type of input will be appropriate.

Version 8.0 of GCG (1994) introduced a graphical interface called WPI

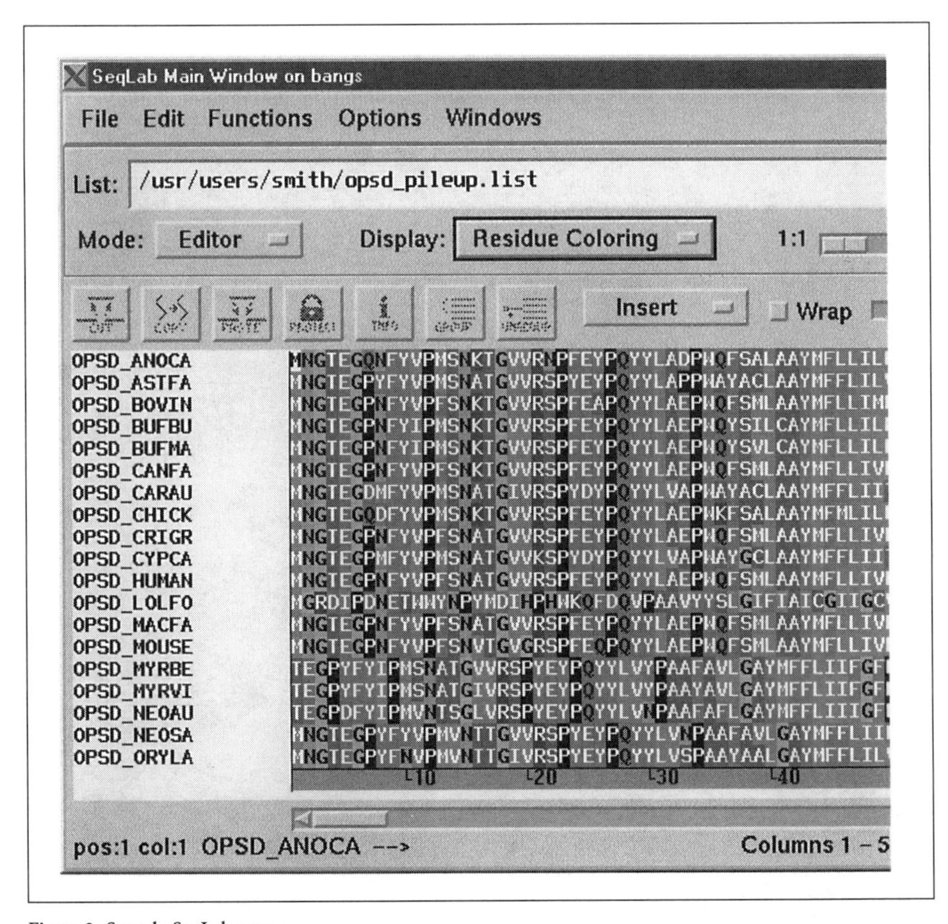

Figure 2. Sample SeqLab screen.

(Wisconsin Package Interface) that made use of the X-Windows communication protocol that provided a menu-driven clickable interface for running GCG programs. However, it was not widely adopted by users. GCG version 9.0 (1997) introduced SeqLab, a vastly improved X-Windows interface that includes a very powerful multiple sequence editor [derived from Steven Smith's GDE (Genetic Data Environment) program (5)] (Figure 2).

GCG has also introduced a Web interface (SeqWeb) for the package in 1998, but at least initially, this has not provided access to all GCG programs and options (Figure 3). There is still the potential for the development of new

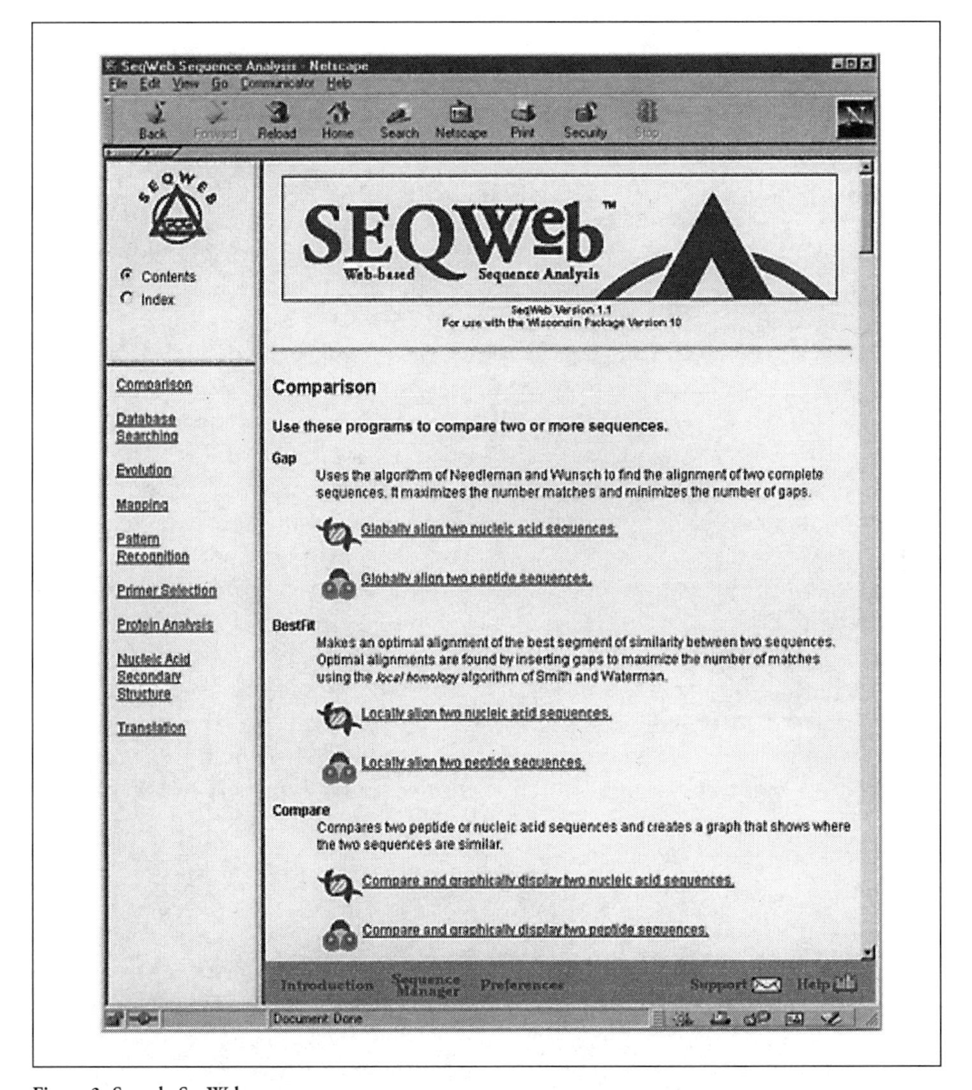

Figure 3. Sample SeqWeb screen.

software that would seamlessly join the Web interface for GCG, ENTREZ, and other databases and molecular biology programs into a single, coherent Web-based molecular biology analysis package.

Output from most GCG programs are simple text files that can be used as input for a wide variety of other programs or formatted in a word processor. Graphical output is available, but requires a rather complex and tedious series of steps. For most types of analyses, GCG has one program to perform the computation and a second to format the results of the computation graphically. It is also possible to transfer graphical output from the mainframe to a PC using a file transfer protocol (FTP) program and then reformat the graphics using a variety of image processing programs on the personal computer.

Several organizations are now providing access to GCG over the Web (SeqWeb) as a completely outsourced solution. Small companies and investigators at universities that do not provide high-quality bioinformatics support can purchase access to GCG from these third-party application service providers less expensively than they can purchase their own copy of GCG (and the computer to run it, the staff to maintain it, and the experts to provide training and technical support to the users). GCG has recently entered a strategic partnership with Viaken Systems (Gaithersburg, MD, USA; **www.viaken.com**) to provide secure Internet access to SeqWeb. Viaken may be able to offer better service and technical support than some university computing centers that provide an out-of-the box GCG installation and out-of-date local databases with minimal technical support from a systems manager who has no knowledge of molecular biology or bioinformatics.

The Australian National Genomic Information Service (ANGIS) has spun off a private company called eBioinformatics Ltd. (**www.ebioinformatics.com**) to provide commercial bioinformatics services to academic and corporate scientists. eBioinformatics offers access to GCG, as well as its own comprehensive suite of Web-based bioinformatics tools called BioNavigator™. BioNavigator tools include PCR primer design, restriction mapping, protein secondary structure prediction, database similarity searching (and scheduled automatic searches), pairwise and multiple alignment, and construction of phylogenetic trees. BioNavigator also provides protocols, step-by-step tutorials that guide users through common bioinformatics procedures such as making multiple alignments or running a similarity search. BioNavigator memberships cost $480 academic/$3000 commercial per user, and additional fees are based on computing time used and amounts of data stored.

REFERENCES

1. **Baran, P.** 1964. On Distributed Communications Networks. IEEE Trans. Comm. Systems.
2. **Cerf, V.G. and R.E. Kahn.** 1974. A protocol for packet network interconnection. IEEE Trans. Comm. Tech., Vol. COM-22. *5*:627-641.
3. **Devereux, J., P. Haeberli, and O. Smithies.** 1984. A comprehensive set of sequence analysis programs for the VAX. Nucleic Acids Res. *12*:387-395.

4. **Metcalfe, R.M. and D.R. Boggs.** 1976. Ethernet: distributed packet switching for local computer networks. Commun. ACM *19*:395-404.
5. **Smith, S.W., R. Overbeek, C.R. Woese, W. Gilbert, and P.M. Gillevet.** 1994. The genetic data environment and expandable GUI for multiple sequence analysis. Comput. Appl. Biosci. *10*:671-675.
6. **Staden, R.** 1980. A new computer method for the storage and manipulation of DNA gel reading data. Nucleic Acids Res. *8*:3673-3694.
7. Wisconsin Package, Version 10.1. 2000. Genetics Computer Group, Madison, WI.

Section II

Computer Tools for Sequence Analysis

4 Finding and Retrieving Sequences in Databases

PUBLIC DATABASES

The public databases of DNA and protein sequences are huge—and growing at an ever increasing rate (1). In order for these databases to be useful, the data must be readily accessible to researchers. Over the years, the methods that have been used to store, distribute, and access this data have changed dramatically.

In the 1970s, Walter Goad, a biochemist at the U.S. Department of Energy (DOE) Los Alamos National Laboratory (LANL), established the Los Alamos Sequence Database as a computerized repository for the primary DNA sequences of genetic material from all organisms reported in the literature (4). The DOE was already involved in biology and biophysics as an outgrowth of its post-WWII studies of the biological effects of irradiation created by atomic weapons.

In 1982, DOE entered into an agreement with the National Institutes of Health (NIH) whereby the Los Alamos Sequence Database was renamed GenBank® and operated as a national repository of DNA sequence information (2). LANL continued to expand and build the database in collaboration with the firm Bolt, Beranek, and Newman under funding provided by NIH, National Institute of General Medical Sciences (NIGMS), and other federal agencies. In 1987, the primary contract for the GenBank project was transferred to the private firm IntelliGenetics (Mountain View, CA, USA), but LANL continued to contribute to database design and maintenance.

In 1992, NIH transferred its management control for the GenBank project from NIGMS to the National Center for Biotechnology Information (NCBI) at the National Library of Medicine, but LANL continued to provide assistance in processing direct submissions for NCBI. After 1993, NCBI developed its own capacity for processing direct submissions and took over complete control of GenBank.

In the 1980s and early 1990s, GenBank was distributed on CD-ROM (compact disc read-only memory) disks. These disks then were used by central computing

Bioinformatics
By Stuart M. Brown

facilities or individual researchers as a searchable database. As GenBank grew larger than a single CD, it became necessary to either mount all of the CDs on multiple CD-ROM drives or copy the data onto large hard drives. This forced many investigators to stop maintaining their own copies of GenBank (and other databases) on personal computers and rely more heavily on central computing facilities.

As researchers became increasingly concerned about searching for newly discovered sequences, which took several months to appear on the next version of the GenBank CD-ROMs, it was necessary for each computing facility to update their databases monthly, weekly, or even daily. The manager of each database had to copy the updated data from the central GenBank computer over the Internet using a procedure known as file transfer protocol (FTP). While FTP is a simple procedure that is easily performed on any type of computer, it presented another obstacle for an individual investigator wishing to access sequence data.

The advent of the World Wide Web (WWW) has revolutionized the process of moving data across the Internet. Using the Web, individual investigators can instantly access GenBank and other databases remotely to search for sequences without the need to store the enormous amounts of data on their own computers. The GenBank database at the NCBI Web site is updated daily, and the ENTREZ search tool is both simple to use and powerful. Many central computing facilities still maintain local copies of databases for use in complex sequence similarity searches and other forms of computation, but for simple search and retrieval of sequences, the Web is usually the best solution. There is the occasional problem of slow Internet access or busy servers, but Internet servers (maintained by major institutes such as NCBI and European Molecular Biology Laboratory [EMBL]) are often more reliable than local computer systems (maintained by universities, companies, or individual scientists).

ENTREZ

An excellent Internet tool for sequence retrieval is called ENTREZ at the NCBI Web site (**http://www3.ncbi.nlm.nih.gov/Entrez/**) (5).

The ENTREZ database contains the most up-to-date nucleotide and protein sequences in GenBank (DNA sequences from GenBank, EMBL, and DNA Databank of Japan [DDBJ]; protein sequences from SWISS-PROT, Protein Information Research [PIR], Protein Research Foundation [PRF], Protein Databank [PDB], and GenPept protein sequences translated from GenBank DNA sequences) as well as the MEDLINE/PUBMED literature database maintained by the National Library of Medicine. In addition, ENTREZ now contains links to the 3-D structures of proteins in the PDB database (6).

ENTREZ provides a powerful "intelligent search engine for the GenBank/MEDLINE database." Beyond simply retrieving sequences, ENTREZ provides precomputed lists of "neighbor" relationships between all of its data elements.

In practice, this means that a search for a text term in sequence annotations or in MEDLINE abstracts will find all journal articles, DNA, and protein sequences that mention that term. All articles and sequences found in a search contain links to related articles or related sequences, such as:

- Relationships between DNA or between protein sequences are calculated with the Basic Local Alignment Search Tool (BLAST) algorithm.
- Relationships between journal articles are computed with Medical Subject Headings (MESH) terms.
- Relationships between 3-D protein structures are calculated by the Vendor Alignment Search Tool (VAST) algorithm.
- Links between DNA and protein sequences, and between proteins and 3-D structures, rely on accession numbers.
- Relationships between sequences and MEDLINE articles rely on both shared keywords and the mention of accession numbers in the articles.

These precomputed relationships might include genes in the same multigene family, articles written about genes that have the same function, or other proteins that function in the same biochemical pathway.

This potential for horizontal movement through the linked databases makes ENTREZ really exciting. It allows the researcher to start with only a vague set of keywords or a sequence identified in the laboratory and rapidly access a set of relevant literature and a list of related database sequences.

Despite the intelligent features of ENTREZ, it still can be frustrating to use. If you know a sequence exists in a public database, do not be discouraged if it is not found with a carefully constructed ENTREZ query. If you have difficulty locating a protein, try searching the DNA database, since the DNA will be linked to its protein translation. Accession numbers are particularly frustrating, since each GenBank sequence may have many different accession numbers. It is possible, for example, that a BLAST search will return a match to a sequence, but the accession numbers in the citation for that sequence will not function as valid search terms in ENTREZ. Be persistent, restructure your query in many different ways, try searching different fields, or using different keywords.

If you are starting with a set of keywords or a general search, rather than trying to retrieve a single sequence with a known accession number, it is a good idea to start with a very broad search. This will produce a large number of "hits" in the database. These sequences can be filtered by adding additional keywords to arrive at subsets that contain just the sequences that are most interesting.

Start by typing a few keywords in the Search field at the top of the NCBI homepage (**http://www.ncbi.nlm.nih.gov**). You must also choose which of the ENTREZ databases you wish to search from the pulldown menu. The default is GenBank—the complete DNA database. Other choices include PUBMED (MEDLINE literature), Proteins, Structures (proteins with known structures in PDB), Genomes (complete genomes and genomes with sequencing in progress), OMIM (Online Mendelian Inheritance in Man—human genes and genetic disorders), and LocusLink (named genes in humans and *Drosophila*). Strangely enough, if you click on the link to ENTREZ in the navigation bar above the search area, the database choices are different. Now you are given PUBMED as the default, Protein, Nucleotide, Genome, Structure, and the new choice of PopSet (polymorphic sequences collected from related organisms), but not OMIM or LocusLink. The location and function of this ENTREZ Search field is the same across all of the main pages of the NCBI Web site, but the database choices vary.

Here is an example using ENTREZ to look for a human protein known as BRI, which has been found to be associated with several degenerative brain

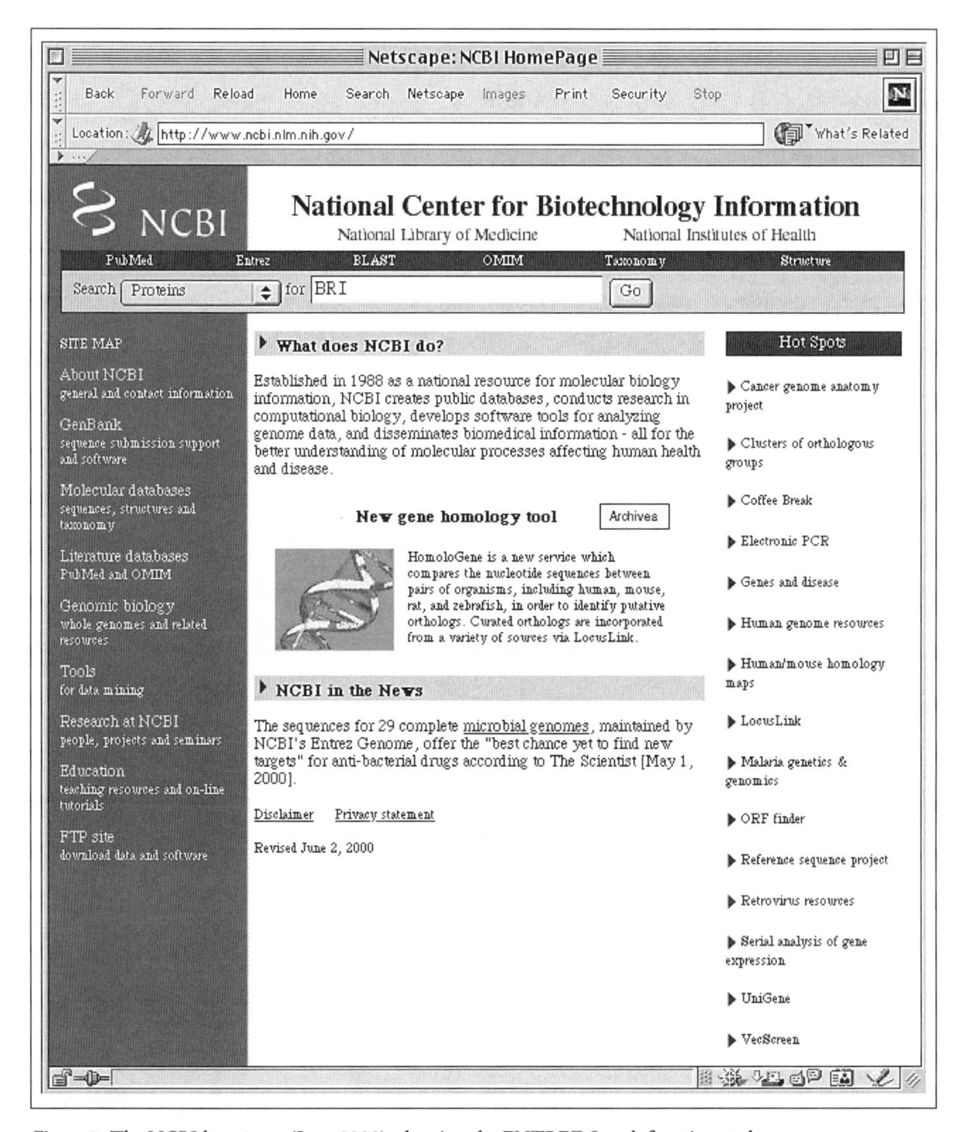

Figure 1. The NCBI homepage (June 2000), showing the ENTREZ Search function at the top.

disorders. For a first try, choose Proteins from the pulldown menu then type BRI in the text entry field and click the Go button (see Figure 1). ENTREZ screens may look somewhat different in your Web browser, either due to differences between computers or due to the constant changes being made at the NCBI.

Nothing is found (see Figure 2). GenBank is like that. It is more a failure of annotation of the sequences than a failure of the ENTREZ search engine. Try searching for the DNA sequence (use Nucleotide or GenBank in the pull-down Search menu depending on what page you are on at the NCBI Web site).

Now you will be shown a rather jumbled collection of genes (see Figure 3), but somewhere in the top 10 should be Accession No. AF200359, the *Homo sapiens* transmembrane protein BRI, which oddly enough is directly linked to a protein accession by a hotlink at the right. If you click on the hotlink, you will see a screen with just that protein accession listed. If you click on the accession number, you will move to a screen that shows the actual protein entry in the GenPept database (see Figure 4).

To the right of the accession number (both in the summary view and in the full GenPept view), there is a link to Related Sequences. If you click on this link, you will see another list of protein sequences (see Figure 5), but this time, they are all related by virtue of significant sequence similarity to the human BRI gene.

Each of the members of this list also has its own links to related protein sequences and to Nucleotides, and some have links to Journal articles in PUBMED. This is where the real power of ENTREZ becomes evident. Each protein sequence is linked to its corresponding DNA sequences, to all similar protein sequences (precomputed with BLAST), and to PDB files if a structure has been determined for that protein. Sequences are also linked to any MEDLINE references that mention that sequence. Each of these sequences and references have

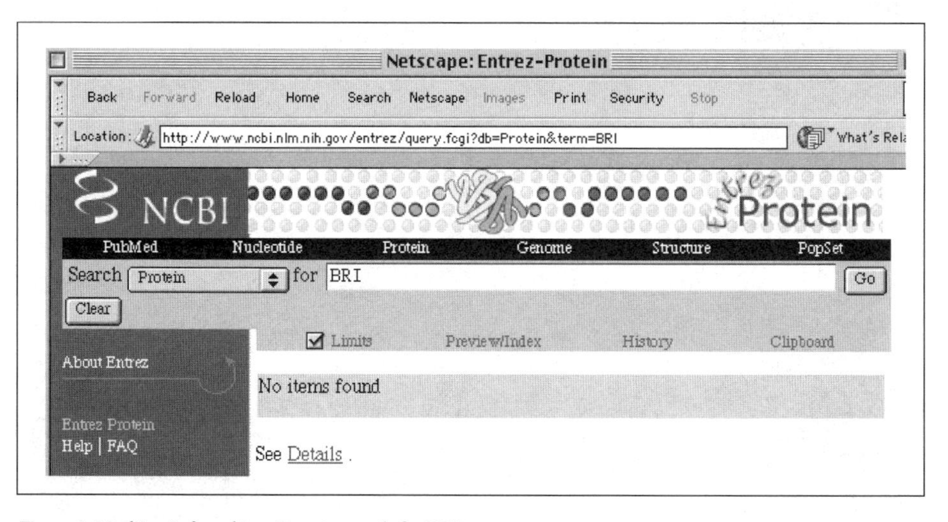

Figure 2. Nothing is found in a Protein search for BRI.

their own links, so from virtually any starting point, you can expand your search horizontally to learn about entire families of related database sequences.

ENTREZ has powerful Boolean logic functions that allow you to build complex search queries using multiple keywords linked by AND, OR, and NOT statements. It tracks a History of all of your searches (in one session), so that you can build complex queries by reusing past search results. It is also possible to Limit a search to specific fields within a document such as Gene Name, Author Name, Page Number, etc., but these features are more useful when searching for Journal Articles in PUBMED rather than for DNA and protein sequences. Remember that the NCBI is part of the National Library of Medicine, so its creators generally think like librarians rather than biologists.

Once a particular GenBank sequence is located with ENTREZ, it is necessary to save that sequence and its annotation to your computer's hard drive. This can be a source of some confusion, since it is not obvious where the information that you are viewing in a Web browser is physically located. It is clear that if you turn off your computer, the information that was in the Web browser window will be

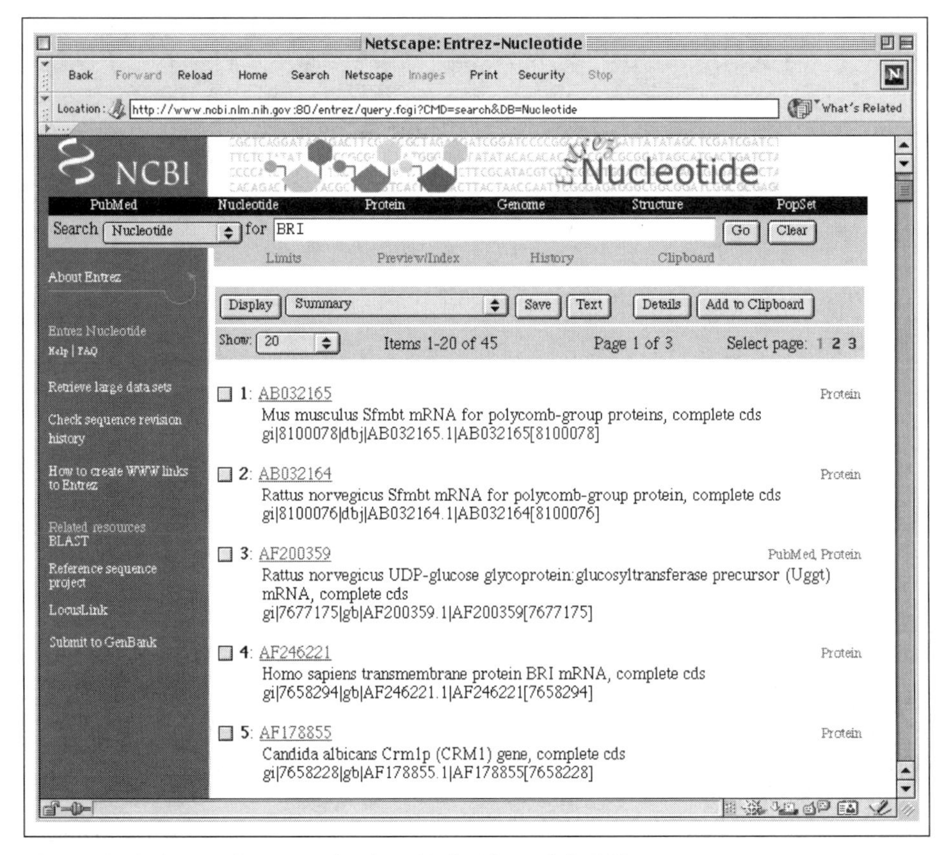

Figure 3. A mixed group of sequences found in a Nucleotide search for BRI.

lost. You can simply copy the information from the screen and paste it into a text file on your computer. Before copying the sequence, it is often advantageous to first change it into FASTA format using the button near the top of the ENTREZ screen, since many other programs will accept data in this format. ENTREZ also provides a Save button at the bottom of the window displaying individual

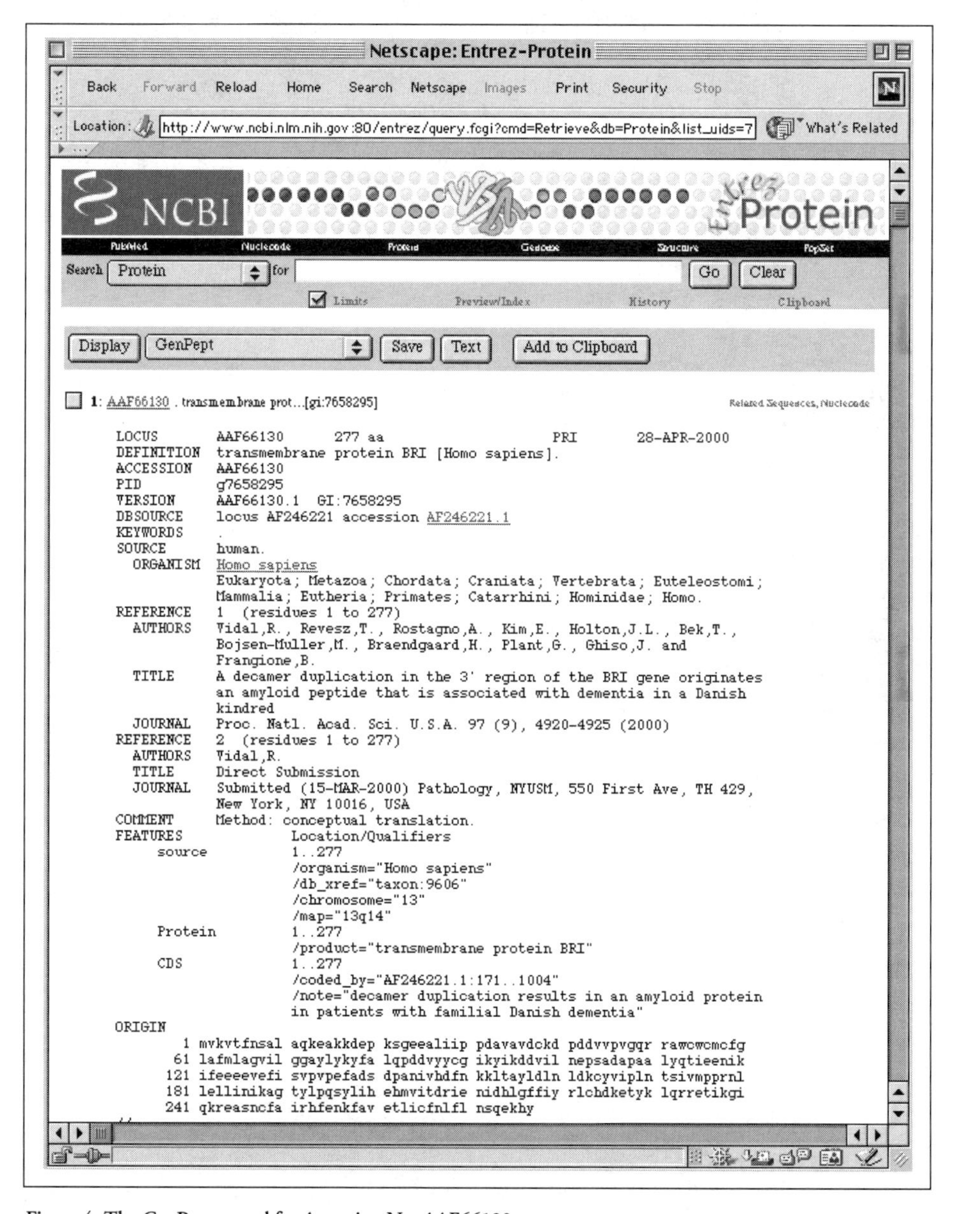

Figure 4. The GenPept record for Accession No. AAF66130.

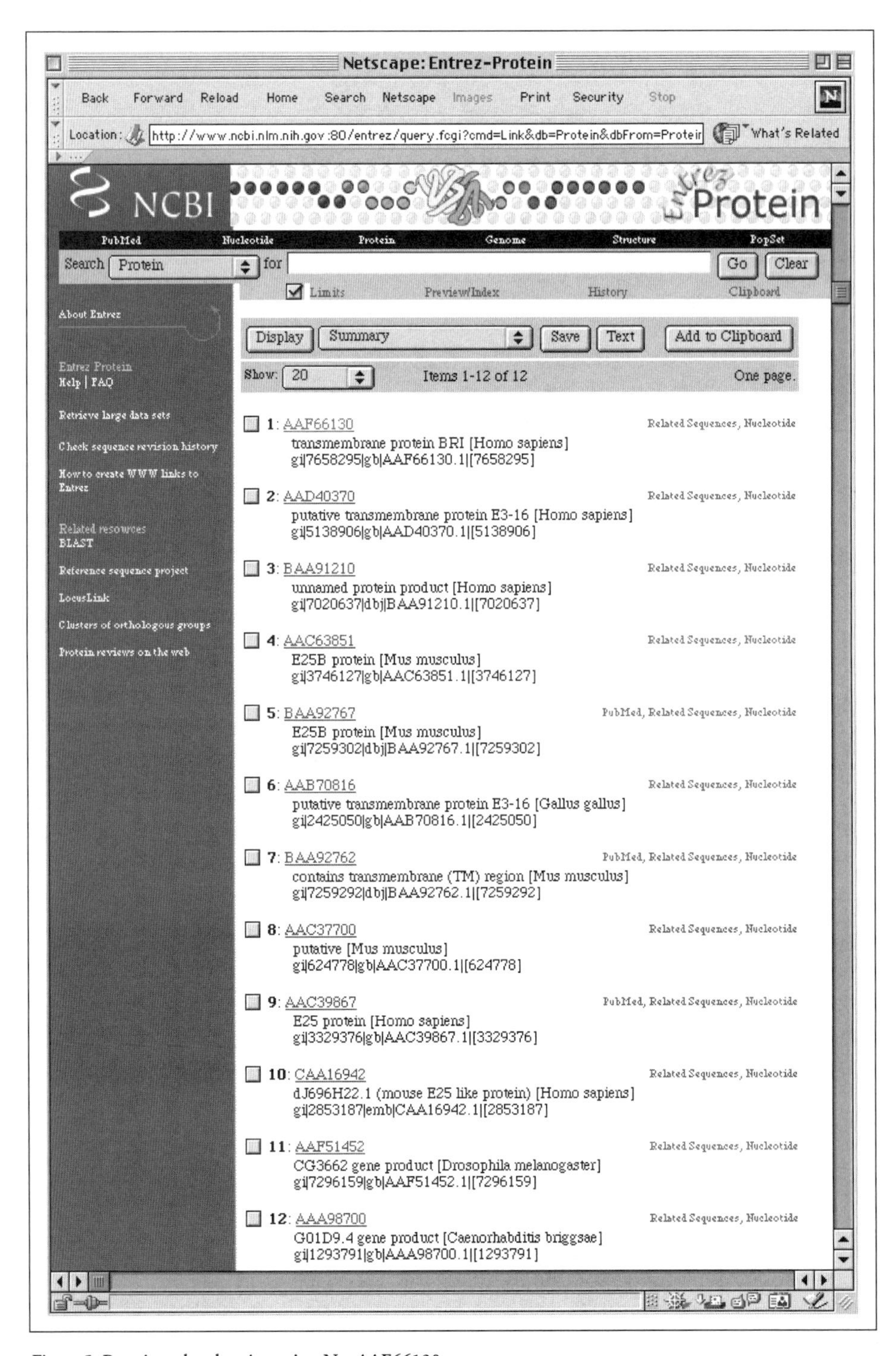

Figure 5. Proteins related to Accession No. AAF66130.

sequence reports, or you can use the Save As... command from your Web browser's menu commands (in the File menu). It is generally simplest to save sequence files as text, so that the sequence itself can be copied and pasted into other sequence analysis programs. However, the file can also be saved as hypertext markup language (HTML). HTML files must be read with a Web browser, however all of the links to other ENTREZ information will still be functional.

SEQUENCE RETRIEVAL SYSTEM

An alternative to ENTREZ is the Sequence Retrieval System (SRS) developed by Thure Etzold and others at the EMBL in Heidelberg (3). SRS is an index of all protein and DNA sequences that can be accessed from many different Web sites. It does not have the added features of MEDLINE journal articles nor the cross referencing of ENTREZ, but it is fast, simple, and not dependent on any single computer or organization. SRS is also available as free software that can be set up on any Web server (or other mainframe machine) to index a set of local databases that can include any public as well as private databases.

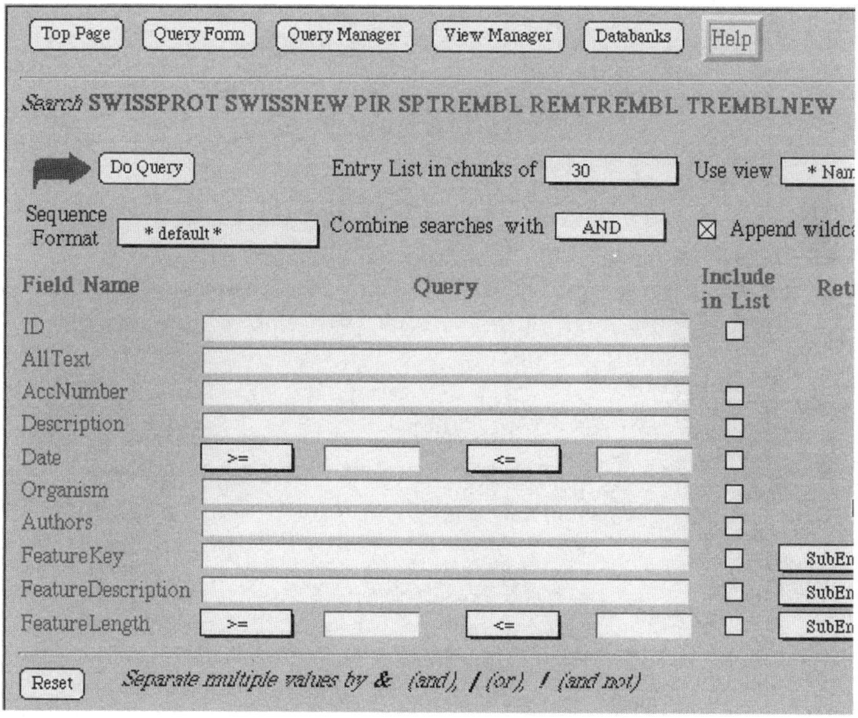

Figure 6. Sample SRS Screen.

GENETICS COMPUTER GROUP LOOKUP AND FETCH

The Genetics Computer Group (GCG) suite of molecular biology programs incorporates a version of SRS as the LOOKUP program. The query window is very similar to SRS on the Web, but it is important to remember that LOOKUP uses a set of database indexes that are created on each local GCG system, so they may not be completely up-to-date, or may not include all databases. The advantage of GCG LOOKUP is that the program produces a text file listing of all sequences that are found by your query. This list file can then be used to automatically retrieve all of the sequences from local databases into your working directory, or even as input for a program that works with sets of sequences such as multiple alignment. This is a big timesaver as compared with cutting and pasting sequences found with SRS or ENTREZ in a Web browser. A sample of the GCG LOOKUP query window follows.

```
Complete the query form below:
                  All text:
                Definition:
                    Author:
                   Keyword:
             Sequence name:
          Accession number:
                  Organism:
                 Reference:
                     Title:
                   Feature:
  On or after (dd-mmm-yy):   On or before (dd-mmm-yy):
 Shortest sequence length:   Longest sequence length:
 Inter-field operator: AND   Form of output list: Whole Entries
Press <Ctrl>Z to continue.
```

REFERENCES

1. **Benson, D.A., M.S. Boguski, D.J. Lipman, J. Ostell, B.F. Ouellette, B.A. Rapp, and D.L. Wheeler.** 1999. GenBank. Nucleic Acids Res. *27*:12-17.
2. **Burks, C., J.W. Fickett, W.B. Goad, M. Kanehisa, F.I. Lewitter, W.P. Rindone, C.D. Swindell, C.S. Tung, and H.S. Bilofsky.** 1985. The GenBank nucleic acid sequence database. Comput. Appl. Biosci. *1*:225-233.
3. **Etzold, T., A. Ulyanov, and P. Argos.** 1996. SRS: information retrieval system for molecular biology data banks. Methods Enzymol. *266*:114-128.
4. **Kanehisa, M., J.W. Fickett, and W.B. Goad.** 1984. A relational database system for the maintenance and verification of the Los Alamos sequence library. Nucleic Acids Res. *12*:149-158.
5. **Schuler, G.D., J.A. Epstein, H. Ohkawa, and J.A. Kans.** 1996. Entrez: molecular biology database and retrieval system. Methods Enzymol. *266*:141-162.
6. **Wang, Y., K.J. Addess, L. Geer, T. Madej, A. Marchler-Bauer, D. Zimmerman, and S.H. Bryant.** 2000. MMDB: 3D structure data in Entrez. Nucleic Acids Res. *28*:243-245.

5 Similarity Searching

INTRODUCTION TO SEQUENCE SIMILARITY

If you have just determined the sequence of an interesting bit of DNA, one of the first questions you are likely to ask yourself is: "Has anybody else seen anything like this?" Fortunately, there has been a very successful international effort to collect all known sequences (both DNA and protein) into databases so they can be searched. The problem is that these databases are huge and, as a result, you must compare your sequence with the vast number of other sequences. A number of computer programs have been written to rapidly compare a query sequence with each of the sequences in a database.

Sequence comparison is the most powerful and most reliable method of answering biological questions about the evolutionary relationships between genes. Recent improvements in statistical methods now allow the biologist to make conclusions with a high level of certainty based on the results of sequence similarity searches. Now that the entire genomes of some organisms have been completely sequenced, it is possible to use similarity search methods to make definitive answers to questions such as "Is there a copy of this interesting gene in this organism?" Or "What genes in this organism have sequences similar to this one?"

A database search is frequently—but incorrectly—referred to as homology searching. The term homology implies a common evolutionary relationship between two traits, whether they are DNA sequences or the bristle patterns on the abdomen of a fly. Just because two sequences share a stretch of nearly identical nucleotides (or amino acids) does not mean that they are directly descended from a common ancestor. Homologous proteins share a common 3-D structure, while proteins with a chance similarity of amino acid sequence do not. Yet over evolutionary time, two homologous proteins may diverge to the point that they do not have any statistically significant similarity, yet retain their common structure. Homologous proteins generally have similar biological functions, but this is not a rigorous requirement—and conversely, discovery of homology is not proof of function.

Bioinformatics
By Stuart M. Brown

In this book, I will call database searching by sequence comparison "similarity searching"; I recommend that you do the same. Of course, a very high level of similarity is a strong indication of homology. As a rule of thumb, 25% identity over a stretch of 100 amino acids can be considered to be good evidence of common ancestry for two sequences. Homologous sequences are usually similar over their entire length, or sometimes just over one functional domain, but it is never correct to state that two sequences are 50% homologous. Homology is an all-or-nothing decision. Be careful in asserting homology between short regions of two sequences; matches that are more than 50% identical in a 20 to 40 amino acid region occur frequently by chance.

All similarity searching methods rely on the concepts of alignment and distance between sequences. A similarity score is calculated from a distance measurement (i.e., the number of DNA bases or amino acids that are different between two sequences), and distances can only be measured between aligned sequences. Therefore, a similarity search boils down to a process of aligning a query sequence with every sequence in a database and calculating some measurement for the quality of each alignment.

Early similarity tools developed by Needelman and Wunch in 1970 (11) and Sellers in 1974 (15) calculated a global similarity score between the entire lengths of the sequences being compared. This type of algorithm is not sensitive for highly diverged sequences; it is better suited for the construction of evolutionary trees than for similarity searching. A better (and faster) method focuses on shorter regions of local similarity (Figure 1). The most widely used local similarity algorithms are Smith-Waterman (16), Basic Local Alignment Search Tool (BLAST) (1), and FASTA (12).

Which program should you use to search a database for similarity to your sequence of interest? This question is almost as controversial as that over choices of computer operating systems (Macintosh® versus Windows or Linux) or religions. In fact, as you enter the world of sequence analysis, you will find religious wars between proponents of different programs over and over. Worse, new programs are constantly appearing.

The Smith-Waterman algorithm is a rigorous dynamic programming approach that is guaranteed to find the optimal local alignments between a query sequence and a database of other sequences. However, Smith-Waterman is not generally used for routine database searching, because it runs very slowly and requires a tremendous amount of computing power. There are some commercially available DNA supercomputers that have been created specifically to speed up the rigorous Smith-Waterman search on huge databases, but this is an expensive option for most researchers, and it has not been proven to be of practical value for most routine similarity analyses.

The molecular biology community has come to rely on shortcut algorithms that use a heuristic approach (based on a process of successive approximations) to perform rapid similarity searches. The techniques used by these rapid heuristic programs result in some loss in the rigor of comparison. It is possible (although, as it turns out, unlikely) that a weak but relevant similarity could be missed by these programs. In addition, many times these programs will identify sequences as being similar to your query sequence when this similarity is not biologically significant. Thus, it is necessary to pay close attention to the statistical significance of similarity results and to exercise careful judgment in the evaluation of the alignments produced by these similarity search programs.

At present, the two most popular heuristic programs for similarity searching are BLAST and FASTA. They use similar approaches to reach similar answers, with some subtle differences—in terms of which sequences are found to be similar, their relative rankings by similarity score, and the statistical significance that is assigned to those similarity scores. For some DNA–DNA searches, FASTA may be more sensitive than BLAST, but BLAST is usually faster. The choice of which program to use is often governed by factors of convenience rather than mathematical rigor; but to be thorough, it is best to perform searches with both programs.

If you are working with DNA sequences, it is worthwhile to remember that DNA is a double-stranded molecule. The protein-coding strand of DNA is always read in the 5′ to 3′ direction (printed left to right by convention), but the complementary strand must be read in the reverse orientation. Both BLAST and

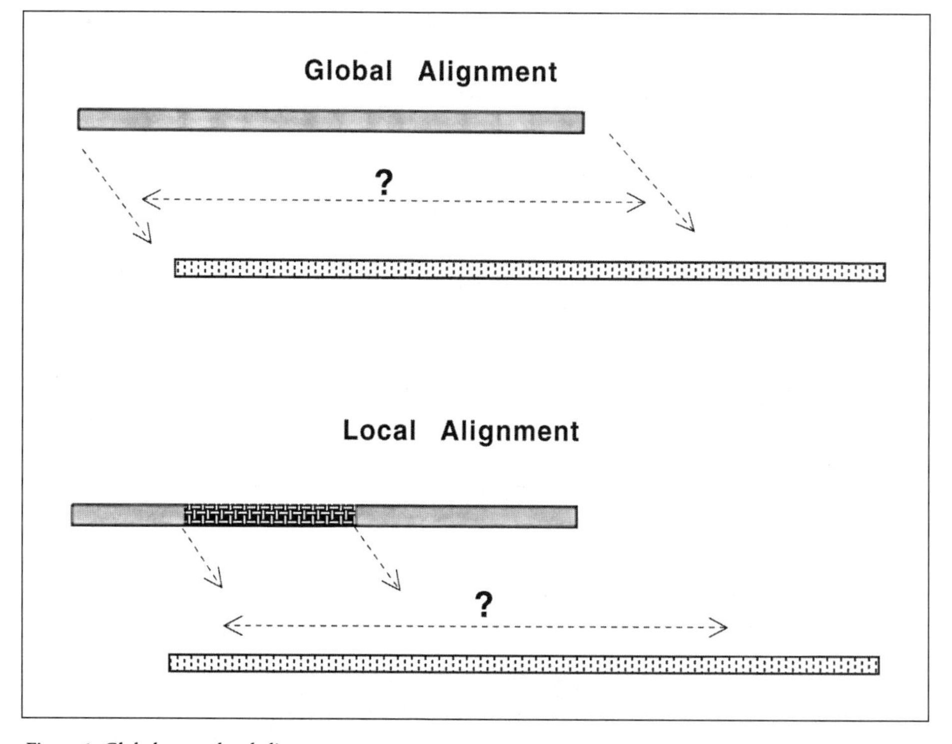

Figure 1. Global versus local alignment.

FASTA automatically search for DNA similarity using both the query sequence that you input and its complementary sequence, thus doubling the computational work. This is important if you are working with an unknown sequence, since you might easily have obtained the sequence of the complementary strand rather than the protein-coding strand. There are also many unknown sequences in the databanks that might be from either strand of the DNA molecule. Sequences of nonprotein-coding regions of DNA are equally likely to be from either strand.

DOT PLOTS

The comparison of two sequences can be done in many different ways. The most direct method is to make this comparison via a visual means in a dot plot. The sequences to be compared are arranged along the axes of a simple graph matrix. At every point in the matrix where the two sequences are identical, a dot is placed (i.e., at the intersection of every row and column that have the same letter in both sequences). A diagonal stretch of dots will indicate regions where the two sequences are similar. Done in this fashion, a dot plot as shown in Figure 2 will be obtained.

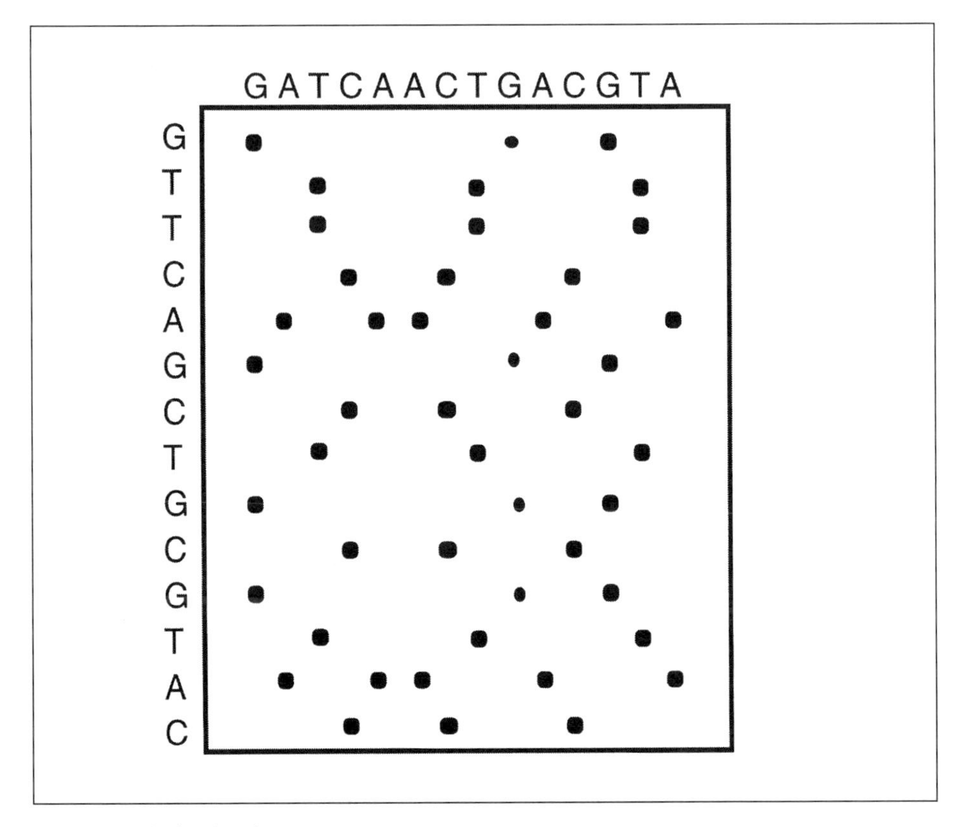

Figure 2. Simple dot plot of two sequences.

A trained eye might be able to pick out a diagonal pattern of similarity in Figure 2, but some statistical methods can be applied to make the results more apparent. Maizel and Lenk (10) popularized the dot plot and suggested the use of a wide variety of filters. The filter recommended by Maizel and Lenk was to create a sliding window and place a dot only when a specified proportion of a small group of bases match. In Figure 3, the same dot plot is shown with a filter such that a dot is printed only if, in a window of 4 bases, 3 of these 4 bases match. To detect more distant similarities, it may be better to use a much larger window (i.e., 20, 30, or even 50 bases) and some suitable percentage of identities (perhaps 50%). However, for real data, these patterns may not be quite so obvious (Figure 4).

DISTANCES

Sequence similarity programs evaluate potential alignments between a pair of sequences based on a calculation of the number of differences between them. The sum of these individual differences is equal to a distance. Distance methods give a single measure of the amount of evolutionary change between two genomes

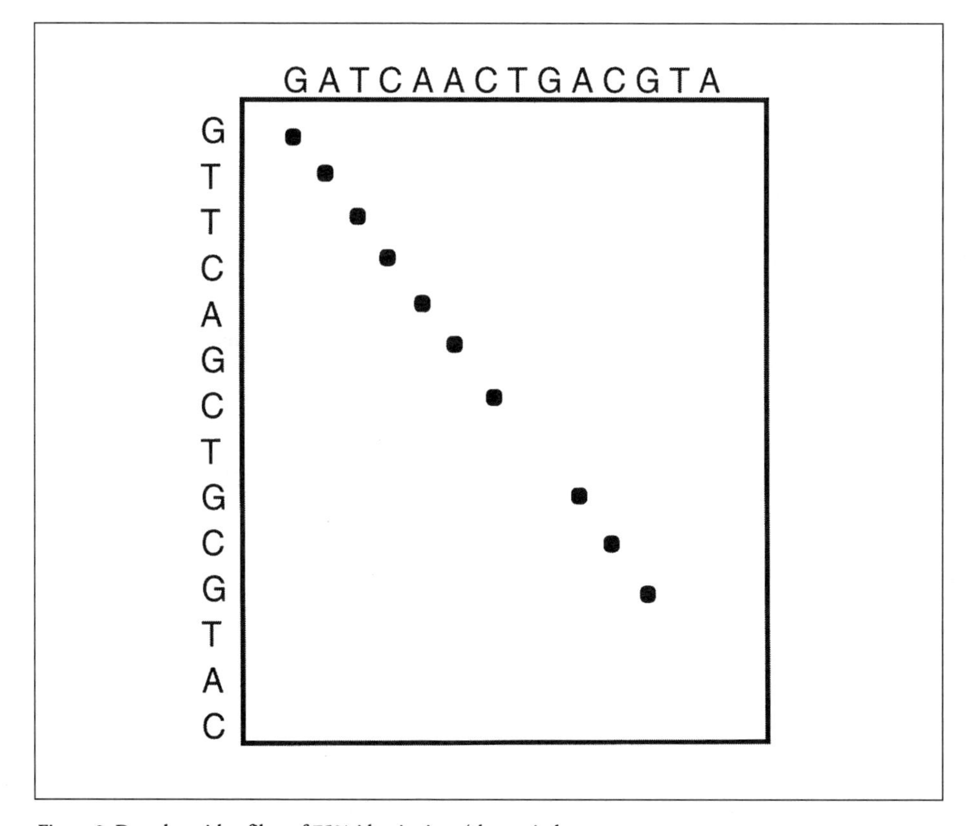

Figure 3. Dot plot with a filter of 75% identity in a 4-base window.

since divergence from a common ancestor. Distances between DNA sequences are relatively simple to compute as a simple sum of the differences between two sequences (this type of algorithm can only work for pairs of sequences that are similar enough to be aligned). Generally, all base changes (mismatches) are considered equally, but a simple matrix of the frequencies of the 12 possible types of replacements (each base can be replaced by one of the three other bases) can be used. Differences due to insertions/deletions (indels) are generally given a larger weight than replacements, but indels of multiple bases at one position are given less weight than multiple independent indels. It is also possible to correct for multiple substitutions at a single site, which is more common in distant relationships and for rapidly evolving sites.

Distances between amino acid sequences are a bit more complicated to calculate. From a functional standpoint, some amino acids can replace one another with relatively little effect on the structure and function of the final protein, while other replacements can be devastating. From the standpoint of the genetic code, some amino acid changes can be made due to the replacement of a single DNA base, while others require two or even three changes in the DNA sequence.

In practice, what has been done is to calculate the frequencies of all amino acid replacements within sets of related amino acid sequences (protein families)

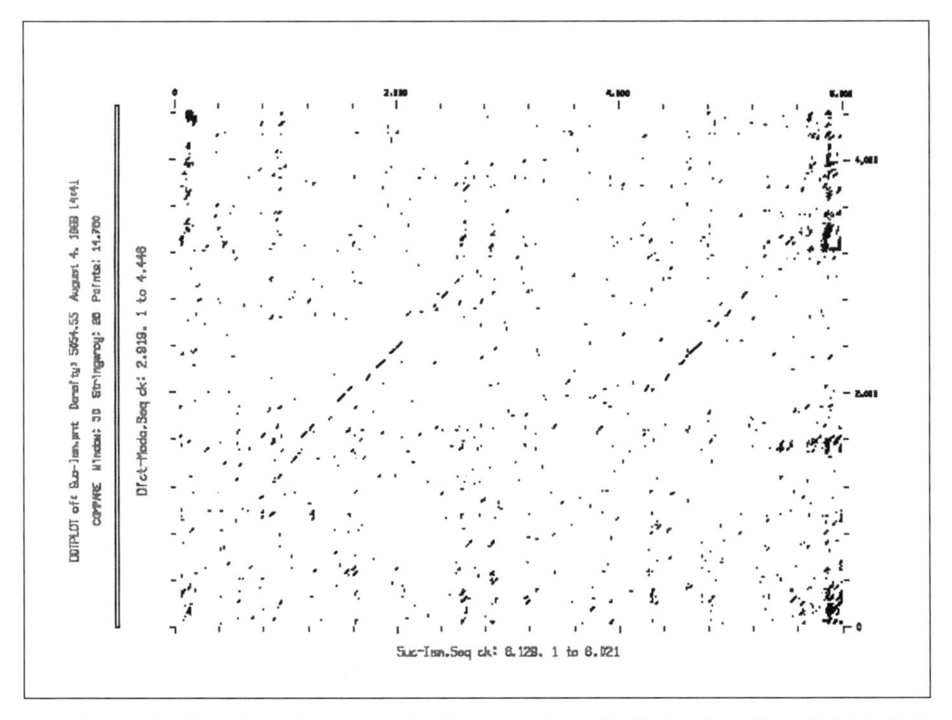

Figure 4. Dot plot of two GenBank sequences that share two regions of similarity using a filter of 54% similarity within a 50-nucleotide window.

in the databanks. The most famous of these is the PAM 250 matrix, created by Margaret Dayhoff in 1978 (5). The PAM stands for "percent accept mutations", also know as a mutation probability matrix, that is, the probability that any amino acid will change to any other amino acid. The 250 in PAM 250 stands for the data gathered from 71 sets of aligned sequences extrapolated up to the level of 250 amino acid replacements per 100 residues. This may seem confusing or impossible, but makes sense in biological terms. Some locations in a protein are much more prone to mutations than others, so that after 250 cycles of mutation, some sites will have mutated several times and some mutations may even reverse themselves, but a 100 amino acid sequence will still have considerable similarity to its original sequence.

The actual PAM 250 matrix (Figure 5) is the log, base 10, of the probability of change from each amino acid residue to any other amino acid. A score of 0 in this matrix indicates that a pair of amino acids replace each other at the frequency predicted by chance occurrence. A score above 0 indicates that these amino acids replace each other more often than expected by chance. That is, they are functionally equivalent and/or easily intermutable. Scores below 0 indicate two amino acids that are seldom interchanged. Interestingly, replacement of each amino acid by itself does not give a constant value. Some amino acids are rare in proteins or

	A	R	N	D	C	Q	E	G	H	I	L	K	M	F	P	S	T	W	Y	V
A	2																			
R	-2	6																		
N	0	0	2																	
D	0	-1	2	4																
C	-2	-4	-4	-5	4															
Q	0	1	1	2	-5	4														
E	0	-1	1	3	-5	2	4													
G	1	-3	0	1	-3	-1	0	5												
H	-1	2	2	1	-3	3	1	-2	6											
I	-1	-2	-2	-2	-2	-2	-2	-3	-2	5										
L	-2	-3	-3	-4	-6	-2	-3	-4	-2	2	6									
K	-1	3	1	0	-5	1	0	-2	0	-2	-3	5								
M	-1	0	-2	-3	-5	-1	-2	-3	-2	2	4	0	6							
F	-4	-4	-4	-6	-4	-5	-5	-5	-2	1	2	-5	0	9						
P	1	0	-1	-1	-3	0	-1	-1	0	-2	-3	-1	-2	-5	6					
S	1	0	1	0	0	-1	0	1	-1	-1	-3	0	-2	-3	1	3				
T	1	-1	0	0	-2	-1	0	0	-1	0	-2	0	-1	-2	0	1	3			
W	-6	2	-4	-7	-8	-5	-7	-7	-3	-5	-2	-3	-4	0	-6	-2	-5	17		
Y	-3	-4	-2	-4	0	-4	-4	-5	0	-1	-1	-4	-2	7	-5	-3	-3	0	10	
V	0	-2	-2	-2	-2	-2	-2	-1	-2	4	2	-2	2	-1	-1	-1	0	-6	-2	4

Figure 5. The PAM 250 scoring matrix (5).

have unusual chemical properties, so their conservation at a particular position is more meaningful than the conservation of a common amino acid. Furthermore, a match between two rare amino acids is extremely unlikely to occur by chance.

A number of refinements have been suggested to the PAM matrix in the years since Dayhoff et al. published this work in 1978 (5). Many of these newer matrixes are optimized for either larger or smaller evolutionary distances than PAM 250. Certain matrixes work better for comparisons within certain protein families. It has not been proven that one particular matrix is best for all comparisons, so it is best to try several different scoring matrixes when comparing protein sequences.

In the 1997 version of Genetics Computer Group (GCG) (v. 9.0), the default scoring matrix for protein comparisons and alignments was changed from PAM 250 to BLOSUM62. The BLOSUM matrix is derived from substitutions observed in more than 2000 blocks of aligned sequences (8), a much larger data set than the one used for the PAM matrix.

SIMILARITY SEARCHES ON THE WORLD WIDE WEB

There are generally three different methods by which researchers can make use of the FASTA and BLAST programs: *(i)* on a PC, *(ii)* on a local workstation or mainframe machine, or *(iii)* over the Internet.

It is generally not practical to use a PC for similarity searching of huge data sets such as GenBank®. The search would take too long (many hours to days depending on the speed of your machine), the data would take up too much hard drive space (18 gigabytes for GenBank in October, 1999), and it is too much work to maintain a current copy of the data (GenBank adds thousands of sequences per day). However, a copy of FASTA on a PC is an efficient way to work with small data sets such as sequences from a personal sequencing project or sequences from one organism. A local supercomputer dedicated to similarity searching would be ideal, but university or company shared computing facilities are generally adequate. For most researchers, the Web has become the best method to make similarity searches. Web servers are available to everyone, they have the most current data, they are usually fast, and your computer is not tied up while your search is computing. However, the addresses and the services offered by various Web servers are subject to change without warning, so it is wise to have some alternatives lined up when your computing is dependent on the kindness of strangers.

BLAST

The National Center for Biotechnology Information (NCBI) BLAST server—fast, free BLAST searches of GenBank.
(**http://www.ncbi.nlm.nih.gov/BLAST/**)

FASTA

Institut de Génétique Humaine, Montpellier, France, GeneStream server.
(**http://www2.igh.cnrs.fr/bin/fasta-guess.cgi**)
Oak Ridge National Laboratory GenQuest server.
(**http://avalon.epm.ornl.gov/**)
European Bioinformatics Institute, Cambridge, England, UK.
(**http://www.ebi.ac.uk/htbin/fasta.py?request**)
European Molecular Biology Laboratory (EMBL), Heidelberg, Germany.
(**http://www.embl-heidelberg.de/cgi/fasta-wrapper-free**)
Munich Information Center for Protein Sequences (MIPS) at Max-Planck-Institut, Germany.
(**http://speedy.mips.biochem.mpg.de/mips/programs/fasta.html**)
Institute of Biology and Chemistry of Proteins, Lyon, France.
(**http://www.ibcp.fr/serv_main.html**)
Institute Pasteur, France.
(**http://central.pasteur.fr/seqanal/interfaces/fasta.html**)
GenQuest at The Johns Hopkins University.
(**http://www.bis.med.jhmi.edu/Dan/gq/gq.form.html**)
National University of Singapore, BioInformatics and BioComputing Server.
(**http://www.cc.um.edu.my/biocomp/fasta.html**)
(results returned by e-mail)
National Cancer Center of Japan.
(**http://bioinfo.ncc.go.jp**)
(results returned by e-mail)
DNA Information and Stock Center, National Institute of Agrobiological Resources, Tsukuba, Japan.
(**http://www.dna.affrc.go.jp/htdocs/homology/homology.html**)

Smith-Waterman

The BIOCCELERATOR at the Weizmann Institute of Science, Israel.
(**http://sgbcd.weizmann.ac.il/genweb**)
GeneStream II at the Institut de Génétique Humaine, CNRS, Montpellier, France. (**http://www2.igh.cnrs.fr/genquest.html**)
(results returned by e-mail)
Oak Ridge National Laboratory.
(**http://avalon.epm.ornl.gov/Grail-bin/GenquestForm_post**)
(results returned by e-mail)

DATABASES

The choice of computer system to use for a search may be influenced by which databases (and subsections of databases) that system has available for searching. GenBank is the largest and most comprehensive DNA database. The EMBL and DNA DataBank of Japan (DDBJ) now contain functionally identical sequence data to GenBank. Sequences are updated between these three organizations daily. Users are advised to use whichever database is closest. The newest entries to GenBank are kept in a separate section that may be known as "gb_new" or "month". Investigators who are anxious to learn of the publication of any new sequence that is similar to the sequences on which they are working may wish to search the new section on a regular basis.

For some searches, it is important to find every possible database match to a given sequence; for others, it may be advantageous to limit the search to just humans, or to mammals, or to the animal kingdom. The measurement of statistical significance in similarity searches is dependent on the size of the database. As the database gets bigger, the chance increases of finding random matches that have similarity scores as large as those of distant homologs, i.e., more noise. The more you can restrict your search, the faster it will run, the better your significance scores will be, and the fewer false-positive hits you will have to sort through (Table 1).

If you are searching DNA databases, another important consideration is whether to search the expressed sequence tags (ESTs), sequence tagged sites (STSs), and the new genomic survey sequence (GSS) and high-throughput genomic sequence (HTGS) databases. These genome project mass sequencing databases are very large, often unannotated, and contain a lot of low-quality, single-pass sequence data. By leaving them out, your search will go significantly faster and leave you with many fewer false matches to sort through. If you are starting with an essentially unknown sequence, then a match against an unannotated EST or genome fragment will probably not contribute much useful information. Once you have completed a search of the well-documented sections of the databases, you might run a separate search against the EST, GSS, and HTGS sections.

There are more choices for protein database searching. The **SWISS-PROT** database, administered by Dr. Amos Bairoch and colleagues at the University of Geneva, Switzerland, is the definitive resource for proteins of known function.

> "SWISS-PROT is a curated protein sequence database which strives to provide a high level of annotations (such as the description of the function of a protein, its domains structure, posttranslational modifications, variants, etc.), a minimal level of redundancy, and high level of integration with other databases" (3).

GenPept is a comprehensive collection of protein translations of all GenBank sequences that have an open reading frame. Many of these amino acid sequences are described as putative proteins with no additional annotation. A match between a new unknown sequence determined in the laboratory and a putative protein in GenPept will not be very helpful in determining the function of the new sequence.

TREMBL is a collection of peptide sequences translated from the EMBL database. It is similar to GenPept, but all SWISS-PROT sequences have been

Table 1. GenBank Searches

GenBank Subdivisions	
gbbct	Bacterial
gbinv	Invertebrate
gbmam	Other mammalian (nonrodent, nonprimate)
gbvrt	Other vertebrate (nonmammalian vertebrates)
gbpat	Patents
gbphg	Phage
gbpln	Plant
gbpri	Primate
gbrod	Rodent
gbsyn	Synthetic sequences (recombinant constructs, etc.)
gbun[a]	Unannotated
gbvrl	Viral
gbest[a]	ESTs (short cDNAs)
gbsts	STSs
gbgss	GSSs (large genomic contigs)
gbhtg	HTGSs (unannotated single-pass sequences produced by the genome projects)

[a]The gb_est section now contains over half of the total data in GenBank. It is further subdivided into subsections gb_est1 to gb_est38 with new sections added with each new release.

removed as well as many redundant sequences. TREMBL plus SWISS-PROT represents nearly the same collection of protein sequences as GenPept.

PDB (Protein Data Bank), maintained by the Brookhaven National Laboratory, is an archive of experimentally determined 3-D structures of biological macromolecules. The archives contain atomic coordinates, bibliographic citations, primary sequence and secondary structure information, as well as crystallographic structure factors and nuclear magnetic resonance (NMR) experimental data.

PIR (Protein Information Resource) is maintained by the U.S. National Biomedical Research Foundation (NBRF) with collaborative support from the Martinsried Institute for Protein Sequences (MIPS) in Germany and the Japanese International Protein Information Database (JIPID) in Japan. The PIR subsection called NRL_3D contains all of the sequence and annotation information in the PDB of known crystallographic structures. PIR has recently (mid-1999) added a new section called PATCHX, which contains a nonredundant set of all known protein sequences that are not contained in the other PIR sections.

OWL, at Johns Hopkins University, is a nonredundant protein sequence database produced from the following source databases: SWISS-PROT, GenPept, PIR, and PDB.

FASTA

FASTA, developed by Dr. William Pearson (13), uses an algorithm that is based on the same concept as a dot plot. In a dot plot, regions of similarity between two sequences show up as diagonals. In FASTA, the sum of the dots along each diagonal is calculated. However, instead of actually constructing a dot plot matrix and then adding up the diagonals, FASTA uses a word-based method. That is, it makes a list of all words in each sequence, where words are (usually) 1 or 2 amino acids long (or 4–6 nucleotides long). This word size is called the *ktup value.* It then identifies words that are identical between the two sequences and checks if these identical words are located near other identical words (i.e., forming a diagonal in a dot plot), but it only counts nonoverlapping words. This is also known to computer programmers as hashing or using a lookup table.

For protein–protein comparisons, regions of the two sequences that share a significant number of identical words are then re-evaluated using a similarity matrix such as the PAM 250. Larger diagonals are created that include similar as well as identical amino acids. Similarity scores are calculated for each diagonal as the sum of similar dots minus mismatches, and the best of these is saved as the value *init1.*

FASTA then tries to join together high-scoring diagonals by adding gaps. New similarity scores are calculated for these diagonals including a gap penalty that may vary depending on the size of the gap. The best score from that is called *initn.* The initn score is saved for each comparison of the query sequence with a database sequence. After all database sequences have been tested, the sequences that produce the best initn scores are used to produce local alignments using a variant of the Smith-Waterman algorithm.

To summarize, FASTA uses four steps to calculate protein similarity scores:

1. Identify regions shared by the two sequences that contain the highest density of single residue identities (ktup = 1) or two consecutive identities (ktup = 2). An approximation of this process can be visualized as a dot plot of the two sequences, which is scanned for diagonals.
2. Rescan the top 10 regions identified in step 1 with the PAM 250 amino acid substitution matrix, trimming the ends so that the regions extend only as long as extending them improves the similarity score. The single best score is stored as init1 for reporting later.
3. Determine if gaps can be used to join the regions identified in step 2, using a penalty of 20 for each gap employed. If so, determine a similarity score for the gapped alignment, which is reported as initn.
4. Construct an optimal alignment of the query sequence and the library sequence, limiting the window to 32 residues centered on the best initial region found in step 2. This score is reported as the optimized score (opt).

FASTA calculates an E-value, which is the expected number of matches that would be found by chance with this same significance score if the database did not actually contain any sequences related to the query sequence. According to

Dr. Pearson, if it is <0.02, the similarity bears further examination; if not, statistical significance is simply not there.

The final output plots the initial scores of each library sequence in a histogram ranked by the z-score, which is derived from the opt score corrected for differences in sequence length (Figure 6a). The general idea of this graph is to show a normal curve of z-scores and E-values that allows you to see the typical values of these statistics for random matches versus the more significant matches at the very bottom of the graph. The graph also shows the distribution of actual alignment scores in comparison with the theoretically expected scores (shown by the "*" marks). If there is not a fairly good match between theoretical and observed scores, then there may be some problem with the search—such as the presence of repeated elements or low complexity sequence in the query.

A list of the most significant scores follows the histogram (Figure 6b), and then the optimal alignments are displayed (the cutoffs for the number of scores and alignments to be shown can be set by the user). The list of matches also contains, for each database sequence, the beginning and end positions of the region of significant similarity to your query sequence. It is also possible to force FASTA to show global alignments between the best hits and your query sequence rather than the local alignments used in its similarity calculations.

FASTA (on proteins) is most sensitive with a word size (ktup) of 1, but the default is 2. Conversely, it will take longer to run a search with a word size of 1 than with one of 2. Similarly, FASTA offers the option of just optimizing the best hits versus the default of optimizing all hits. With optimization on all hits, FASTA runs somewhat slower, but is able to catch more distant similarities (14).

FASTA is available as part of the GCG package or as a stand-alone program for Macintosh, PC, UNIX®, and most other computer platforms. The entire GenBank database would take a very long time to search with FASTA on a Mac or PC, but FASTA is a convenient tool for working with smaller personal databases, since it works with data in a simple format. One major advantage of using GCG is that it is possible to use the results of a FASTA search as input for a PILEUP multiple alignment or for other GCG programs that use lists of sequences.

FASTA FORMAT

FASTA format is a compact and simple method of storing DNA and protein sequences as text files that can be read by virtually all molecular biology programs including (finally!) GCG versions 9.0 and up.

A sequence in FASTA format begins with a single-line description (or header), followed by lines of sequence data. The description line is distinguished from the sequence data by a greater-than symbol (>) in the first column. It is recommended that all lines of text be shorter than 80 characters in length. An example sequence in FASTA format is shown in Figure 7.

Sequences are represented in the standard IUB/IUPAC amino acid and nucleic

```
Histogram Key:
Each histogram symbol represents 143 search set sequences
Each inset symbol represents 3 search set sequences
z-scores computed from opt scores

z-score obs    exp
        (=)    (*)

< 20     261      0 :*======
  22      74      0 :*=
  24      95      0 :*==
  26     133      1 :*===
  28     103      6 :*==
  30     158     35 :*===
  32     197    135 :===*=
  34     321    367 :========*
  36     576    754 :===============   *
  38     934   1246 :=======================       *
  40    1444   1738 :====================================      *
  42    1853   2124 :==============================================     *
  44    2372   2343 :=========================================================*
  46    2402   2387 :==========================================================*
  48    2428   2285 :=======================================================*====
  50    2287   2085 :==================================================*=====
  52    1985   1833 :============================================*====
  54    1643   1566 :======================================*==
  56    1366   1308 :===============================*==
  58    1165   1074 :==========================*==
  60     851    870 :=====================*
  62     665    697 :================*
  64     535    555 :=============*
  66     394    438 :==========*
  68     309    345 :========*
  70     241    270 :======*
  72     196    211 :=====*
  74     162    165 :====*
  76     117    128 :===*
  78      83    100 :==*
  80      81     77 :=*
  82      43     59 :=*
  84      37     47 :=*
  86      23     36 :*
  88      14     28 :*
  90      13     22 :*
  92      13     17 :*          :=============   *
  94       3     13 :*          :===         *
  96       7     10 :*          :=======  *
  98       6      8 :*          :====== *
 100       1      6 :*          :=    *
 102       3      5 :*          :=== *
 104       3      4 :*          :===*
 106       0      3 :*          :  *
 108       1      2 :*          :=*
 110       0      2 :*          : *
 112       2      1 :*          :*=
 114       0      1 :*          :*
 116       0      1 :*          :*
 118       0      1 :*          :*
>120      15      0 :*          *===============

Results sorted and z-values calculated from opt score
1619 scores saved that exceeded 66;  22182 optimizations performed
Joining threshold: 46, optimization threshold: 34, opt. width: 32
```

Figure 6a. FASTA search using the TREMBL prokaryote database using sp:P43738 RPOB_HAEIN (*Haemophilus influenzae* rpoB) histogram.

```
The best scores are:                          init1 initn   opt    z-sc E(25408)..

Trembl_Pro:P94850      Begin: 8  End:  880
! P94850 helicobacter pylori (campylo...  650  2384  2572  2655.6        0
Trembl_Pro:O08406      Begin: 69  End:  847
Trembl_Pro:O08406      Begin: 69  End:  847
! O08406 mycobacterium tuberculosis. ... 1297  3682  2297  2368.6        0
Trembl_Pro:Q59564      Begin: 75  End:  852
! Q59564 mycobacterium tuberculosis. ... 1242  3467  2258  2328.7        0
Trembl_Pro:Q50388      Begin: 43  End:  843
! Q50388 mycobacterium smegmatis. dna... 1242  3425  2210  2278.4        0
Trembl_Pro:Q49003      Begin: 24  End:  344
! Q49003 mycoplasma capricolum. dna-d...  733  1178  1162  1201.5        0
Trembl_Pro:Q59554      Begin: 1  End:  124
! Q59554 mycobacterium smegmatis. rna...  633   633   638   666.4  6.6e-31
Trembl_Pro:Q53458      Begin: 1  End:  98
! Q53458 neisseria meningitidis. rna ...  597   597   600   628.8  8.2e-29
Trembl_Pro:Q54230      Begin: 21  End:  177
! Q54230 streptomyces griseus. rna-po...  215   378   346   361.3  6.6e-14
Trembl_Pro:Q48975      Begin: 110  End:  249
! Q48975 mycoplasma capricolum. dna-d...  187   385   349   359.4  8.3e-14
Trembl_Pro:P72031      Begin: 2  End:  57
! P72031 mycobacterium tuberculosis. ...  313   313   316   338.5  1.2e-12
Trembl_Pro:Q49439      Begin: 1  End:  82
! Q49439 mycoplasma genitalium. uncer...  307   452   307   323.6  8.3e-12
Trembl_Pro:Q49441      Begin: 47  End:  112
! Q49441 mycoplasma genitalium. uncer...  243   243   247   262.1  2.2e-08
Trembl_Pro:Q59056      Begin: 39  End:  414
! Q59056 methanococcus jannaschii. me...   55    91   141   140.3     0.13
Trembl_Pro:P76307      Begin: 9  End:  103
! P76307 escherichia coli. from bases...   67    67   116   126.4      0.8
Trembl_Pro:Q59156      Begin: 447  End:  667
! Q59156 anaerocellum thermophilum. d...   45    45   127   122.7      1.3
\\End of List

                     10        20        30        40        50        60
Hin1637.Pep  MGYSYSEKKRIRKDFGKRPQVLNVPYLLTIQLDSFDKFIQKDPEGQQGLEAAFRSVFPIV
                    |:|:| || | |  |:|| || :| ||:|:|: :    ::|:| :|:|:|||
P94850       MSKKIPLKNRLRADFTKTPTDLEVPNLLLLQRDSYDSFLYSKDGKESGIEKVFKSIFPIQ
                     10        20        30        40        50        60

                     70        80        90       100       110
Hin1637.Pep  SNNGYTELQYVDYRLEEPEFDVRECQIRGSTYAAGLRVKLRLVSYDKESSS--RAVKDI
             ::::   |:|:  :: : :: ||| : || ||:  |::|:||: ::|:: |    ::|||
P94850       DEHNRITLEYAGCEFGKSKYTVREAMERGITYSIPLKIKVRLILWEKDTRSGEKNGIKDI
                     70        80        90       100       110       120

             120       130       140       150       160       170
Hin1637.Pep  KENEVYMGEIPLMTDNGTFVINGTERVIVSQLHRSPGVFFDSDKGKTHSSGKVLYNARII
             ||: ::: ||||||:  :|:|||:|||:|:||||||||:|  ::::| | :|::|:::||
P94850       KEQSIFIREIPLMTERTSFIINGVERVVVNQLHRSPGVIFKEEESST-SLNKLIYTGQII
                    130       140       150       160       170
```

Figure 6b. FASTA search using the TREMBL prokaryote database using sp:P43738 RPOB_HAEIN (*H. influenzae* rpoB) list of best scores and part of the first alignment.

```
>gi|532319|pir|TVFV2E|TVFV2E envelope protein
ELRLRYCAPAGFALLKCNDADYDGFKTNCSNVSVVHCTNLMNTTVTTGLLLNGSYSENRT
QIWQKHRTSNDSALILLNKHYNLTVTCKRPGNKTVLPVTIMAGLVFHSQKYNLRLRQAWC
HFPSNWKGAWKEVKEEIVNLPKERYRGTNDPKRIFFQRQWGDPETANLWFNCHGEFFYCK
MDWFLNYLNNLTVDADHNECKNTSGTKSGNKRAPGPCVQRTYVACHIRSVIIWLETISKK
TYAPPREGHLECTSTVTGMTVELNYIPKNRTNVTLSPQIESIWAAELDRYKLVEITPIGF
APTEVRRYTGGHERQKRVPFVXXXXXXXXXXXXXXXXXXXXXXVQSQHLLAGILQQQKNL
LAAVEAQQQMLKLTIWGVK
```

Figure 7. TVFV2E sequence in FASTA format.

acid codes, with these exceptions: *(i)* lowercase letters are accepted and are mapped into uppercase; *(ii)* a single hyphen or dash (–) can be used to represent a gap of indeterminate length; and *(iii)* in amino acid sequences, U and * are acceptable letters. Before submitting a request, any numerical digits in the query sequence should either be removed or replaced by appropriate letter codes (e.g., N for unknown nucleic acid residue or X for unknown amino acid residue).

BLAST

Another very popular searching algorithm is BLAST(1). The NCBI (in Washington, DC) offers free BLAST searches of GenBank over the Web at (**http://www.ncbi.nlm.nih.gov/BLAST/**). This service is (usually) remarkably fast despite heavy use by scientists throughout the world. The NCBI's BLAST computers have a tremendous amount of random access memory (RAM) and so are able to keep entire databases in active memory at one time, and thus do not have to read it in from disk for each search. This is probably one reason why BLAST is so fast.

BLAST is actually a family of programs for searching various combinations of protein and DNA sequences:

1. BLASTP compares an amino acid query sequence against a protein sequence database.
2. BLASTN compares a nucleotide query sequence against a nucleotide sequence database.
3. BLASTX compares a nucleotide query sequence translated in all reading frames against a protein sequence database.
4. TBLASTN compares a protein query sequence against a nucleotide sequence database dynamically translated in all reading frames.
5. TBLASTX compares the six-frame translations of a nucleotide query sequence against the six-frame translations of a nucleotide sequence database. (This is a very computation-intensive calculation, not allowed on NCBI's BLAST Web server for the entire nonredundant GenBank database).

BLAST is available as a component of the GCG package, and it is also available as a stand-alone program for Mac, PC, and all major types of mainframe computers. GCG can be configured to use BLAST on a local database or to access GenBank at the NCBI via the Internet. BLAST requires databases in its own special format, so it uses a significant amount of hard disk space to keep a local copy of GenBank and the other major databases in BLAST format. It is possible to create a BLAST database with GCG, so one can make a BLAST search using one's own personal database of sequences.

Like FASTA, BLAST is also a word-based method. However, the method is considerably different from that used in FASTA. BLAST takes the first word from the query sequence (for proteins, a word is 3 amino acids long, for DNA 11 nucleotides) and then locates all of the similar words in the first database sequence. A similar word is defined by a score greater than a certain cutoff value, calculated using an amino acid comparison matrix such as BLOSUM62 (DNA–DNA word matches are based only on the percentage of identical bases). BLAST does not require identical matches between words in the query and test sequences. Any database sequence that does not have a significant word match with the query sequence is removed from further consideration.

For each pair of similar words, BLAST tries to build an alignment between the query and test sequence locations by expanding the initial word match in both directions. Since BLAST does not allow gaps, it only expands each alignment until adding additional residues no longer increases the similarity score for the segment pair. It saves the best similarity score for each alignment in the query sequence. At the end of this process, it selects a set of high-scoring segment pairs (HSPs) that represent the similarity scores for the best alignments. It then moves on to the next test sequence. At the end of the database search, it has a list of the largest HSPs and returns that list as the output of the search. It also attempts to combine several HSPs that are in order and nonoverlapping for purposes of improving the similarity score, but it will not build a gapped alignment with these segments. You will see these in the output file as a series of segments marked with P(N), where N is 2, 3, or whatever (Figure 8).

In late 1997, the NCBI introduced BLAST 2.0, which offers two new options, Gapped BLAST and Position Specific Iterated-BLAST (PSI-BLAST) (2). These new options represent significant changes in the BLAST algorithm, but the upshot for the researcher is more sensitivity to weak but biologically meaningful sequence similarities.

Gapped BLAST allows the introduction of gaps (deletions and insertions) into alignments. Gapped BLAST produces longer continuous alignments rather than the multiple short segments in the basic BLAST output. Also, the scoring of gapped results tends to be more biologically meaningful than ungapped results. The inability of BLAST to utilize gaps in alignments has long been its major weakness, so this new algorithm promises to be a major improvement. In addition to improving sensitivity, generating longer aligned regions, and improving the predictive power of the statistical scores, gapped BLAST is about three times faster than the traditional BLAST algorithm. This is very important since the rate of growth of the databases is now faster than the rate of improvement in computer processor speeds.

The basic BLAST algorithm compares two sequences by breaking them into sets of short words, and then looks for close matches between pairs of words between two sequences. Wherever statistically significant matches are found between word pairs, BLAST tries to create an alignment by extending the match in both directions. The new gapped BLAST algorithm requires matches between two different pairs of words located near each other on the two sequences before it tries to create an alignment. Since the chance of randomly finding two different matching words between two unrelated sequences is much lower than the chance for a single random match, the gapped BLAST algorithm is able to use a less stringent cutoff score

```
BLASTP 1.4.9MP [26-March-1996] [Build 14:27:01 Apr  1 1996]

Query=  tmpseq_1
        (1343 letters)

Database:  Non-redundant SwissProt sequences
           68,619 sequences; 24,728,842 total letters.
Searching..................................................done

                                                                   Smallest
                                                                     Sum
                                                            High  Probability
Sequences producing High-scoring Segment Pairs:             Score  P(N)      N

sp|P43738|RPOB_HAEIN DNA-DIRECTED RNA POLYMERASE BETA CHA...  6637  0.0        1
sp|P06173|RPOB_SALTY DNA-DIRECTED RNA POLYMERASE BETA CHA...  2949  0.0        2
sp|P00575|RPOB_ECOLI DNA-DIRECTED RNA POLYMERASE BETA CHA...  2936  0.0        2
sp|P41184|RPOB_BUCAP DNA-DIRECTED RNA POLYMERASE BETA CHA...  2750  0.0        2
sp|P19175|RPOB_PSEPU DNA-DIRECTED RNA POLYMERASE BETA CHA...  1338  0.0        6
sp|P77941|RPOB_RICTY DNA-DIRECTED RNA POLYMERASE BETA CHA...  1196  0.0        8
sp|P30760|RPOB_MYCLE DNA-DIRECTED RNA POLYMERASE BETA CHA...   959  0.0       12
sp|P47766|RPOB_MYCTU DNA-DIRECTED RNA POLYMERASE BETA CHA...   920  0.0       11
sp|Q59622|RPOB_NEIME DNA-DIRECTED RNA POLYMERASE BETA CHA...   908  0.0       10
sp|P77965|RPOB_SYNY3 DNA-DIRECTED RNA POLYMERASE BETA CHA...   871  0.0        7
.
.
sp|P23579|RPOB_EUGGR DNA-DIRECTED RNA POLYMERASE BETA CHAIN    381  5.3e-228 13
sp|Q46124|RPOB_CAMJE DNA-DIRECTED RNA POLYMERASE BETA CHA...   583  2.9e-213  4
sp|Q51561|RPOB_PSEAE DNA-DIRECTED RNA POLYMERASE BETA CHA...   611  1.5e-124  2
sp|P08036|RPOB_SAPOF DNA-DIRECTED RNA POLYMERASE BETA CHAIN    349  9.8e-112  5
sp|P50546|RPOB_ARATH DNA-DIRECTED RNA POLYMERASE BETA CHAIN    314  1.7e-98   3
sp|P31814|RPOB_THECE DNA-DIRECTED RNA POLYMERASE SUBUNIT B     232  1.7e-75  11
sp|P11513|RPOB_SULAC DNA-DIRECTED RNA POLYMERASE SUBUNIT B     197  3.4e-70  12
sp|Q03587|RPOB_THEAC DNA-DIRECTED RNA POLYMERASE SUBUNIT B     220  1.6e-69  10
sp|P21421|RPOB_PLAFA DNA-DIRECTED RNA POLYMERASE BETA CHAIN    266  2.8e-66   6
sp|Q10578|RPB2_CAEEL DNA-DIRECTED RNA POLYMERASE II SECON...   195  4.1e-58  11
sp|P30876|RPB2_HUMAN DNA-DIRECTED RNA POLYMERASE II 140 K...   208  2.3e-57  12
sp|P22703|RPOB_ANASP DNA-DIRECTED RNA POLYMERASE BETA CHA...   452  4.0e-55   1
sp|P08518|RPB2_YEAST DNA-DIRECTED RNA POLYMERASE II 140 K...   203  5.0e-55  11
.
.
sp|P05472|RPOL_KLULA PROBABLE DNA-DIRECTED RNA POLYMERASE...   146  2.8e-23   5
sp|P41187|RPOB_LIBAF DNA-DIRECTED RNA POLYMERASE BETA CHA...   149  1.9e-20   2
sp|P09844|RPB2_METTH DNA-DIRECTED RNA POLYMERASE SUBUNIT B"    128  1.4e-17   6
sp|P15352|RPB2_HALHA DNA-DIRECTED RNA POLYMERASE SUBUNIT B"    126  2.8e-17   5
sp|P41558|RPB2_METVA DNA-DIRECTED RNA POLYMERASE SUBUNIT B"    129  6.1e-17   7
sp|Q58444|RPB2_METJA DNA-DIRECTED RNA POLYMERASE SUBUNIT B"    130  1.2e-14   5
sp|P14657|DHE4_UNKP  NADP-SPECIFIC GLUTAMATE DEHYDROGENAS...    56  0.23      6
sp|P19450|ACSC_ACEXY CELLULOSE SYNTHASE OPERON C PROTEIN        71  0.77      1
sp|P77938|NADC_RHORU PROBABLE NICOTINATE-NUCLEOTIDE PYROP...    69  0.90      1
sp|P30765|RS10_MYCLE 30S RIBOSOMAL PROTEIN S10                  54  0.998     2

sp|P06173|RPOB_SALTY DNA-DIRECTED RNA POLYMERASE BETA CHAIN
            (TRANSCRIPTASE BETA CHAIN) (RNA POLYMERASE BETA SUBUNIT)
            Length = 1342

Query:   181 GSWLDFEFDPKDNLFARIDRRRKLPATIILRALGYTTEEILNLFFDKITFEIAGDKLLMT 240
             GSWLDFEFDPKDNLF RIDRRRKLPATIILRAL YTTE+IL+LFF+K+ FEI  +KL M
Sbjct:   181 GSWLDFEFDPKDNLFVRIDRRRKLPATIILRALNYTTEQILDLFFEKVVFEIRDNKLQME 240

Query:   241 LVPERLRGETASFDIEANGKVYVERGRRITARHIKALEKDNISQVVVPSEYILGKVASKD 300
             L+PERLRGETASFDIEANGKVYVE+GRRITARHI+ LEKD+I  + VP EYI GKV SKD
Sbjct:   241 LIPERLRGETASFDIEANGKVYVEKGRRITARHIRQLEKDDIKHIEVPVEYIAGKVVSKD 300

Query:   301 YVDLESGEIICPANGEISLETLAKLAQAGYTTIETLFTNDLDYGPYISETLRVDPTYDKT 360
             YVD  +GE+IC AN E+SL+ LAKL+Q+G+  IETLFTNDLD+GPYISET+RVDPT D+
Sbjct:   301 YVDESTGELICAANMELSLDLLAKLSQSGHKRIETLFTNDLDHGPYISETVRVDPTNDRL 360

Query:   361 SALYEIYRMMRPGEPPTPESSEALFNNLFFSAERYDLSTVGRMKFNRSLAFPEGEGAGIL 420
             SAL EIYRMMRPGEPPT E++E+LF NLFFS +RYDLS VGRMKFNRSL   E EG+GIL
Sbjct:   361 SALVEIYRMMRPGEPPTREAAESLFENLFFSEDRYDLSAVGRMKFNRSLLRDEIEGSGIL 420
```

Figure 8. BLASTP search of the SWISS-PROT database using sp:P43738 RPOB_HAEIN (*H. influenzae* rpoB); abridged list of best score and partial alignment to sp:P06173 (RPOB_SALTY).

for considering two similar words to be a match. The net result is a more sensitive search, with less computer time wasted trying to extend uninformative random word matches, and also better, longer, and more meaningful alignments.

PSI-BLAST uses a technique similar to GCG's PROFILESEARCH. PSI-BLAST first performs an initial gapped BLAST search of the database. Then, it makes a multiple alignment from the significant hits. The multiple alignment is then used to construct a position-specific scoring matrix—a table of amino acid frequencies at each position in the sequence. This matrix is then used instead of a query sequence for another BLAST search. New alignments found in the second search are incorporated into the scoring matrix, and the process is repeated (iterated) until no more significant hits are found. Each cycle of BLAST searching with a matrix takes only as long as a simple BLAST search, and the process of building a multiple alignment and calculating the matrix is very fast compared to the time spent on the database search.

PSI-BLAST is able to find highly diverged members of protein families. This is quite similar to (but much faster and easier than) the procedure that most investigators usually follow of collecting meaningful hits from a BLAST (or FASTA) search, building a multiple alignment with PILEUP, then using PROFILEMAKE and PROFILESEARCH to search for additional protein sequences that are similar to the consensus of the multiple alignment. PSI-BLAST also makes obsolete another quick and dirty technique that many researchers use, which is to use each of the hits from one database search as query sequences for additional searches and then compile a list of all sequences that are found in multiple searches. This remarkable PSI-BLAST program is very new and still in its evaluation phase, but it could have a major impact on the process of identifying the function of new genes as they are sequenced by the Genome Projects.

One drawback of the PSI-BLAST program is that it is dependent on the results of an initial BLAST search with a single query sequence to build the first multiple alignment and searching matrix. If you start with a protein that has no significant similarity to anything else in the database, then no amount of matrix building and repeated searching will produce more significant matches. Furthermore, if you start the search with a sequence that is part of a highly diverged branch of a large protein family, then the initial matrix will be skewed toward the members of that branch of the family (the closest relatives of your query sequence), and more distantly related proteins may not be found regardless of the number of iterations.

The databases available for BLAST searching (at NCBI) are shown below. If you are using BLAST on a GCG system that is configured with local databases at your site, the database selection may be different.

Peptide (Protein) Sequence Databases

nr = All nonredundant GenBank CDS translations + PDB + SWISS-PROT + PDB.

month = All new or revised GenBank CDS translation + PDB + SWISS-PROT + PDB released in the last 30 days.
SWISS-PROT = The SWISS-PROT protein sequence database.
yeast = Yeast (*Saccharomyces cerevisiae*) protein sequences.
PDB = Sequences derived from the 3-D structure Brookhaven Protein Data Bank.
kabat = Kabat's database of sequences of immunological interest.
alu = Translations of select Alu repeats.

Nucleotide Sequence Databases

nr = All nonredundant GenBank + EMBL + DDBJ + PDB sequences (but no ESTs or STSs).
month = All new or revised GenBank + EMBL + DDBJ + PDB sequences released in the last 30 days.
dbest = ESTs.
dbsts = STSs.
yeast = Yeast (*Saccharomyces cerevisiae*) genomic nucleotide sequences.
PDB = Nucleotide sequences derived from 3-D protein structures in the Brookhaven Protein Data Bank.
kabat = Kabat's database of sequences of immunological interest.
vector = Vector subset of GenBank.
mito = Database of mitochondrial sequences, Rel. 1.0, July 1995.
alu = Select Alu repeats.
epd = Eukaryotic promoter database.
gss = Genome survey sequence, includes single-pass genomic data, exon-trapped sequences, and Alu PCR sequences.

BLAST CONSIDERATIONS

BLAST is a very complicated tool if you look at it closely. Here are some things you should keep in mind:

1. The default nonredundant DNA database does not include the EST, STS, GSS, or HTGS databases.
2. BLAST is fast because it initially throws away all database sequences that do not have a significant match without gaps to the query sequence. If two sequences are generally similar over a long region, but do not have a single highly conserved region, BLAST might miss the similarity. Sequences that have many small insertion/deletions throughout a region of high similarity may hide from BLAST.
3. BLAST works much better for proteins than for nucleotides because it requires a high similarity score in a window of 11 nucleotides in order to start an HSP. BLAST can find protein sequences that have diverged by approximately

250 substitutions per 100 amino acids, but only 50 substitutions for 100 nucleotides. If you must make a similarity search with a DNA sequence against a DNA database, then FASTA will be able to find more diverged matches.

4. The output from BLAST can be enormous. You do not want to blindly print it, as it will often run to a couple of hundred pages. Rather, go in with an editor, delete the pieces you don't need, and then print what is left. If you are running BLAST from GCG, you can restrict the number of hits returned with the /LISTSIZE and /SEGMENTS commands. LISTSIZE restricts the summary of hits at the top of the BLAST output, SEGMENTS restricts the number of detailed alignments that follow.
5. BLAST is particularly sensitive to short repeats (microsatellites) and skewed composition (GC- or AT-rich DNA sequences, proline-rich proteins). This is because the model it uses for extending HSPs does not handle these common sequence anomalies correctly. If you leave them in, you will have a large number of bogus matches. You should remove these from your query sequence with the option /Filter=xs (filtering is on by default for the NCBI BLAST Web server). If the filter is applied, repeated and low complexity regions are replaced with XXXs that are ignored by BLAST.
6. If your sequence comes from an organism (or organelle) that does not use the standard genetic code, be sure to use the option /TRANSLATE=N to specify the proper translation table. For instance, yeast mitochondrial is 3, and vertebrate mitochondrial is 2.
7. There are a lot of other parameters you can adjust in the BLAST program that can change the way it behaves. It is a good idea to read the entire BLAST manual to get a better idea of how to use these options.

BLAST VERSUS FASTA

BLAST and FASTA do the same job in different ways. One is not better than the other, and in most cases both should be used to insure the most thorough search. However, there are some considerations that might favor one program over the other for particular applications.

1. FASTA is more sensitive than BLAST for DNA–DNA searches, especially for highly diverged sequences.
2. BLAST is better at finding short regions of high similarity, while FASTA is better at finding long regions of lower similarity. BLAST will miss similar sequences if they do not have a single identical word (11 DNA bases or 3 amino acids).
3. Results from a BLAST search are generally composed of many short aligned regions between two sequences. FASTA produces longer aligned regions. FASTA can provide a full-length alignment between pairs of similar sequences (with the /SHOWALL option).
4. BLAST searches of GenBank are available as a fast, free service over the

Web at (**http://www.ncbi.nlm.nih.gov/BLAST/**).
5. FASTA searches are available over the Web from several sources, but these services change from time to time. Results may be returned by e-mail and may take some time (several hours).
6. In GCG, the output from FASTA searches can immediately be used as a list file for multiple alignment (with PILEUP) or for any other program that can be applied to a list of sequences.
7. FASTA is generally easier to install on a local computer (Mac, PC, UNIX, VMS®) and therefore more control of databases and subsections of databases to be searched is gained.

GENERAL TIPS FOR SIMILARITY SEARCHING

Protein similarity searches can find much more distant similarities than comparisons of DNA sequences (approximately 2.5 billion versus 100 million years of evolutionary divergence). This is true for several reasons. First, the DNA alphabet is restricted to only four letters, while the amino acid alphabet has 20 letters, so the probability of chance matches is much greater with DNA–DNA comparisons. Second, two differing DNA bases can only be scored as a mismatch, while two amino acids can share varying degrees of similarity based on their physical and chemical properties, etc. Third, the protein databanks are much smaller than the DNA databanks, so searches can be more sensitive without incurring too many false positives.

Dr. William Pearson (author of FASTA) says: "The number one thing that you should learn is that in general, you should try not do DNA sequence comparison, unless you are really interested in a sequence that does not code for a protein. Comparing proteins instead of DNA sequences is the most effective way of improving your searches."

Generally, if your query sequence is protein, you will search protein databanks, and with DNA sequence you will search nucleotide data. However, it is possible to automatically translate your DNA sequence into amino acids in all six reading frames (BLASTX and FASTX) and compare it to protein databases, or to compare your protein sequence to the six reading frame translation of all DNA database sequences (TFASTA and TBLASTN). Surprisingly, a protein–protein search using a translated DNA sequence as a query against a protein databank, or a protein query against a translated DNA databank, is much more sensitive than a DNA–DNA search.

Another important concept in similarity searching is the protein family (or gene family). These families can be measured in two dimensions, across species and across related functions. Clearly, closely related species will have similar sets of genes that perform similar functions. Even distantly related species can be expected to have some similar genes that perform basic cellular functions such as protein synthesis and DNA replication. By definition, similar proteins that perform identical functions in different species are called orthologs. There are also families of related proteins within a single organism such as tyrosine kinases or G-protein-coupled receptors. The proteins within these families are clearly homologous, having similar sequences and similar 3-D structures, but they have diverged by a process of gene duplication, mutation, and selection for different functions. Such related proteins are called paralogs. The line between orthologs and paralogs grows less distinct when proteins are compared between distantly related organisms—is a bacterial protein a homologue to a human protein that performs an identical function if the two share 15% sequence identity?

All of these search algorithms will fail to find significant similarity between two homologous sequences (i.e., members of a single protein family) at some distance or evolutionary divergence between them. As illustrated in Figure 9, using sequence A as a query, we may find B but not be able to find C, yet C may really be distantly related to A and a bit more closely related to B. Another way of looking at this is that similarity is transitive. So if A is similar to B, and B is similar to C, then A is similar to C.

The scores reported by similarity programs are called expectation values (E-values), but they should be treated just like the *P* values from traditional types of statistical tests. These expectation values represent the number of times that a match with the observed similarity score would occur by chance if the database actually contained no sequences that were truly similar to your query sequence (i.e., in a database of random sequences). Generally, expectation scores smaller than 0.05 or 0.02 are considered significant. The results of a similarity search can be divided into three categories: *(i)* highly significant matches between nearly identical proteins, *(ii)* insignificant matches that are clearly due to chance similarities, and *(iii)* an intermediate range of matches with high, but not statistically significant, similarity scores.

These intermediate values have been described as the "twilight zone" (6). Some of those twilight-zone proteins are distantly related homologues, while others are unrelated sequences with chance high-similarity scores. One simple method to test for distant homologues is to use each twilight-zone protein as the query for another similarity search. True homologues will have significant matches, not only with your original query sequence, but also to other members of that

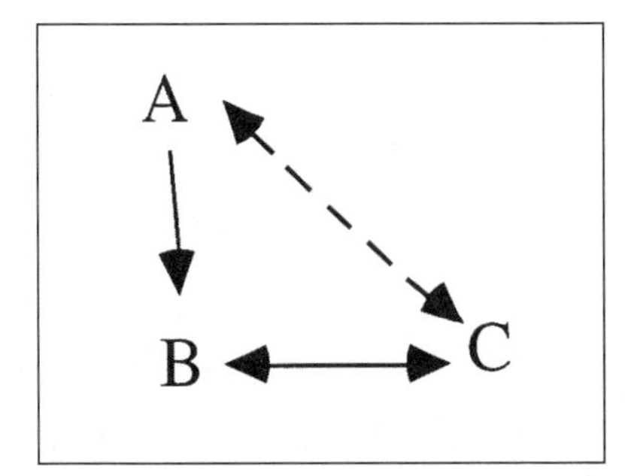

Figure 9. Sequence similarity is transitive; so if A is similar to B, and B is similar to C, then A is also similar to C.

protein family (other sequences that received highly significant scores in the original search). Unrelated sequences are less likely to match other proteins in the family of your original query sequence (14).

A common pattern for the results of a similarity search with BLAST or FASTA is that a small number of closely related sequences will receive highly significant E-values (E < 0.001), but another group of more distantly related sequences will receive E-values that are not clearly significant (E > 0.1). It is often possible to use additional searches to find a bridge between these two groups (14). Therefore, it is important to take the sequences that are identified in an initial database search and to repeat the search with each of them. Obviously, you want to redo searches with the more divergent hits—if the original query was mouse trypsin, there isn't much point repeating the search with rat trypsin!

This process of using the significant matches from one BLAST search as query sequences for other searches has been automated as a Java™ program on the Iterative Neighborhood Cluster Analysis (INCA) Web site (**http://itsa.ucsf.edu/~gram/home/inca/**) (7). INCA runs a BLAST search on a given query sequence. It then runs BLAST on each of the significantly similar sequences found in the first iteration. All of the significant matches from the searches are saved in a composite cluster. A Java-enabled Web browser such as Netscape 4.0 or Internet Explorer 4.0 is required to use INCA.

From a practical perspective, a similarity search tool for comparing a newly sequenced gene to the database should identify evolutionarily related sequences and ignore chance similarities. For a given similarity search program, there is always a tradeoff between sensitivity, for detecting distantly related sequences, versus the number of unrelated false positives that are found in a search. These false positives and false negatives can be sorted out by repeated searches with each questionable sequence looking for family relationships.

OTHER SIMILARITY SEARCH METHODS

FRAMESEARCH and TFASTX

There is a program available in the GCG package called FRAMESEARCH that is specifically designed for dealing with bad DNA sequence data that contains insertions and deletions that cause frameshifts in protein-coding sequences. It can find an alignment between a protein query and a nucleotide test sequence even if the DNA contains frame-shifting gaps (or insertions). There is also a stand-alone program called TFASTX, which is part of the FASTA 3.0 package, that performs a similar gapped alignment of protein query to a DNA databank.

A lot of sequences in the databases, especially the ESTs, have frameshifts. A fair estimate is that over half of the EST sequences have these errors. The usual method for comparing a protein to these EST sequences is to use either the TFASTA or the TBLASTN program, which automatically translates the DNA sequences in all six reading frames, and then makes separate similarity calculations with each translated sequence. Frameshifts break the translated proteins into segments that end up in different reading frames. If the segments are large enough, then some similarity may still be detected by TBLASTN or TFASTA. In a sequence with multiple frameshifts, there might not be a single open reading frame long enough to show up as significant in a TBLASTN or TFASTA search. FRAMESEARCH and TFASTX will spot the frameshifts and create a correctly gapped translated sequence that will give an accurate similarity score and can be used in alignments.

FRAMESEARCH, unfortunately, is a very slow program, as it is an extended version of the Smith-Waterman method. A complete search of the entire GenBank EST database might take more than 24 hours. FRAMESEARCH can be used more efficiently to screen a large list of EST hits found with either FASTA or BLAST. Low-quality matches to the ESTs found by these programs may have short regions of similarity to your query sequence, but frameshifts might be masking larger similar regions that can be revealed with FRAMESEARCH. TFASTX is much more rapid than FRAMESEARCH, since it relies on a version of the FASTA program that is forgiving of gaps, but it is still much slower than a standard DNA–DNA or protein–protein similarity search.

Pattern Matching

Another approach to similarity searching is to look for a specific pattern. This can work in two directions: either to look for a set of known patterns in a query sequence or to search for a query pattern in a large database. Known patterns that might be searched for in a query protein include the conserved protein functional domains in the PROSITE database or more general structural properties such as transmembrane regions, coiled coil regions, or intracellular localization signals (transit peptides). DNA patterns include intron–exon splicing boundaries, promoters, and other transcriptional regulatory regions. Database searches with query patterns can be conducted with conserved regions identified in multiple alignments, or regions from a single protein (or DNA sequence) that are suspected of having functional significance.

Searching a single sequence with a set of known patterns is a simple computational operation. There are a wide variety of programs that can perform this task on a PC and many different Web sites also provide this function—often using unique custom databases of patterns. Conversely, searching an entire database with a test pattern is a very large computational job that generally cannot be done on a PC and is rarely offered (with restrictions) on Web sites. This is the type of heavy work that is best done with GCG running on a fast mainframe machine. GCG has an excellent set of pattern-matching tools including MOTIFS, FINDPATTERNS, and PROFILESEARCH. The subject of pattern matching is discussed in greater detail in Chapter 7.

Pattern matching is generally not as powerful a tool for finding homologous sequences as the similarity algorithms discussed above. By definition, homologous proteins share a common 3-D structure and therefore must have large regions of similar sequence. However, pattern matching can be used advantageously to identify genes or proteins that share short functional domains. While pattern-matching algorithms can readily identify exact matches to a query pattern, they handle mismatches and gaps less intelligently than straightforward similarity algorithms. If a pattern search is done with the parameters set to tolerate too many mismatches, it will indiscriminately find a large number of unrelated sequences.

INTERPRETATION OF OUTPUT

Interpretation of the output from similarity search programs is the most difficult part of the analysis. It is necessary to understand the strengths and weaknesses of the computer program that you are using and integrate the results shown by the program with knowledge of the biology of the genes being analyzed.

BLAST and FASTA produce similar output files (see Figure 6 and 7). First, there is a short description of the program and a list of the databases and program options chosen. Then, there is a list of all of the database sequences that matched your query sequence. A number is assigned to each of these sequences that represents the quality of the match. The list is presented in descending order, so that the best matches are at the top of the list. However, the most biologically significant matches are not always the ones ranked highest in the list.

A quick look at this list of hits can provide a lot of information. Are there any hits with E-scores of zero or smaller than 10^{-100}? These would be nearly exact matches—most likely the same gene in related species or a recently duplicated gene in a single species. Are there lots of hits with highly significant scores and a very gradual decline from the 10^{-100} down to 0.01? This would indicate that your query sequence is a member of a large protein family and would suggest further research in the protein family and protein domain databases and in the literature.

Following the list of hits are the alignments of your query sequence with each of the database sequences (only the region of best alignment is shown). If a database sequence has more than one region with strong similarity to regions of your query sequence, then that sequence will appear more than once in the list, and more than one aligned region will be shown. BLAST groups all regions of alignment between two sequences together, but FASTA lists each alignment in strict rank order based on its matching quality statistic.

It is up to you, the biologist, to scrutinize these alignments and determine if they are significant. Were you looking for a short region of nearly identical sequence or a larger region of general similarity? Are the mismatches conservative ones? Are the matching regions important structural components of the genes or just introns and flanking regions? In general, a short match of 25 amino acids with 50% identity (or even higher for DNA) can be expected just by chance in any search of a large database, while a match of 100 amino acids with 25% identity/similarity is a strong indicator of true homology.

Some Common Mistakes

While it is true that BLAST and FASTA will miss some distant relationships, it is more common that these tools will report matches as significant even though they are not truly homologous. The most common source of these false-positive errors stem from low-complexity (biased composition) or repeated sequences.

The BLAST Web server uses the SEG (18) and DUST (4) filtering programs by default to remove some short-periodicity internal repeats and low-complexity sequences, but many types of repeats get past these filters.

A similarity search using the *Trypanosoma brucei* CRAM protein (GenBank locus TRBCARM2) provides an excellent example of false matches due to both biased sequence composition and repeated sequences. The first match shown in Figure 10 is due simply to the high concentration of aspartic acid in the two proteins. The second alignment shows a match between two repeating patterns; the 12 amino acid repeat in CRAM shares two identical and one similar amino acids with the six amino acid repeat in a *Plasmodium falciparum* surface antigen (AF056936).

The obvious conclusion to be drawn from this example is that the results of similarity search programs must be carefully inspected by humans. Furthermore, a highly significant similarity score is not proof that two sequences are homologous.

```
>24600 (Closest domain: SPBP_RAT 153-279)
 PROSTATIC SPERMINE-BINDING PROTEIN PRECURSOR (SBP) Length = 127
   Score = 93  (40.6 bits), Expect = 0.002
  Identities = 42/134 (31%), Positives = 58/134 (42%), Gaps = 14/134 (10%)

 Query:     13 DDCNITGDCNETDDCDITGDCNETDDCNITGDCNETDDCDITGDCNETDDCNITGDCNET 72
               DD +   D  E DD D   D NE D  +   D +  DD D   D  E D+     D  E
 Sbjct:    154 DDFDDNDDDKEDDD-DEHDDDNEEDHGDKDNDNDHDDDHDDDDDDKEDDNEEDVDD—ER 210

 Query:     73 DDCNITGDCNETDDCNITGDCNETDDCNITGDCNETDDCNITGDCNETDDCNITGDCNET 132
               DD     D +E DD N     N+ DD   +GD ++ DD     + ++ DD   +GD  +
 Sbjct:    211 DD——KDDDEEDDDN—DKENDKDDGEGSGDDDDNDD——-EDDDKDDDGGSGDDGDD 259

 Query:    133 DDCDITGDCNETDD 146
                D D   D  + D+
 Sbjct:    260 GDDDEDDDGGDDDN 273


gi|3044185 (AF056936) mature parasite-infected erythrocyte surface antigen
           [Plasmodium falciparum] Length = 1661
 Score =  106 bits (263), Expect = 1e-22
 Identities = 54/178 (30%), Positives = 90/178 (50%)

Query: 4   NITGDCNETDDCNITGDCNETDDCDITGDCNETDDCNITGDCNETDDCDITGDCNETDDC 63
            N+ G+  ETD+   T +  ET +   TG+  ET +   TG+  ET +   TG+  ET +
Sbjct: 203 NVMGESKETDESKETDESKETGESKETGESKETGESKETGESKETGESKETGESKETGES 262

Query: 64  NITGDCNETDDCNITGDCNETDDCNITGDCNETDDCNITGDCNETDDCNITGDCNETDDC 123
              TG+  ET +   TG+  ET +   TG+  ET +   TG+  ET +   TG+  ET +
Sbjct: 263 KETGESKETGESKETGESKETGESKETGESKETGESKETGESKETGESKETGESKETGES 322

Query: 124 NITGDCNETDDCDITGDCNETDDCNITGDCNETDDCNITGDCNETDDCNITGDCNETE 181
             +T    ET    IT +  ET++  IT +  + +   ++G    +++ N+T +  ET+
Sbjct: 323 KVTRIYEETKYTKITSEFRETENVKITEESKDREGNKVSGPYENSENSNVTSESEETK 380
```

Figure 10. BLAST similarity search results with *T. brucei* CRAM protein (GenBank locus TRBCARM2) illustrating false positives due to low sequence complexity and repetitive patterns.

REFERENCES

1. **Altschul, S.F., W. Gish, W. Miller, E.W. Myers, and D.J. Lipman.** 1990. Basic local alignment search tool. J. Mol. Biol. *215*:403-410.
2. **Bairoch, A and R. Apweiler.** 2000. The SWISS-PROT protein sequence database and its supplement TREMBL in 2000. Nucleic Acids Res. *28*:45-48.
3. **Altschul, S.F., T.L. Madden, A.A. Schäffer, J. Zhang, Z. Zhang, W. Miller, and D.J. Lipman.** 1997. Gapped BLAST and PSI-BLAST: a new generation of protein database search programs. Nucleic Acids Res. *25*:3389-3402.
4. **Claverie, D.M. and D. States.** 1993. Information enhancement methods for large scale sequence analysis. Comput. Chem. *17*:191-201.
5. **Dayhoff, M.O., R.M. Schwartz, and B.C. Orcutt.** 1978. A model of evolutionary change in proteins, matrixes for detecting distant relationships, p. 345-358. *In* M.O. Dayhoff (Ed.), Atlas of Protein Sequence and Structure, Vol. 5. National Biomedical Research Foundation, Washington, DC.
6. **Doolittle, R.F.** 1986. Of ORFs and URFs: A Primer on How to Analyze Derived Amino Acid Sequences. University Science Books, Mill Valley, CA.
7. **Graul, R.C. and W. Sadée.** 1997. Evolutionary relationships among proteins probed by an iterative neighborhood cluster analysis (INCA). Alignment of bacteriorhodopsins with the yeast sequence YRO2. Pharm. Res. *14*:1533-1541.
8. **Henikoff, S. and J.G. Henikoff.** 1992. Amino acid substitution matrices from protein blocks. Proc. Natl. Acad. Sci. USA *89*:10915-10919.
9. **Lipman, D.J. and W.R. Pearson.** 1985. Rapid and sensitive protein similarity searches. Science *227*:1435-1441.
10. **Maizel, J.V., Jr. and R.P. Lenk.** 1981. Enhanced graphic matrix analysis of nucleic acid and protein sequences. Proc. Natl. Acad. Sci. USA *78*:7665-7669.
11. **Needelman,S.B. and C.D. Wunch.** 1970. A general method applicable to the search for similarities in the amino acid sequence of two proteins. J. Mol. Biol. *147*:195-197.
12. **Pearson, W.R.** 1990. Rapid and sensitive sequence comparison with FASTP and FASTA, p. 63-98. *In* R.F. Doolittle (Ed.), Methods Enzymology, Vol. 183. Academic Press, San Diego.
13. **Pearson, W.R.** 1997. Identifying distantly related protein sequences. Comput. Appl. Biosci. *13*:325-332.
14. **Pearson, W.R. and D.J. Lipman.** 1988. Improved tools for biological sequence comparison. Proc. Natl. Acad. Sci. USA *85*:2444-2448.
15. **Sellers.** 1974. SIAM J. Appl. Math. *26*:787-793.
16. **Smith, T.F. and M.S. Waterman.** 1981. Identification of common molecular subsequences. J. Mol. Biol. *147*:195-197.
17. **Wilbur, W.J. and D.J. Lipman.** 1983. Rapid similarity searches of nucleic acid and protein data banks. Proc. Natl. Acad. Sci. USA *80*:726-730.
18. **Wootton, J.C. and S. Federhen.** 1996. Analysis of compositionally biased regions in sequence databases. Methods Enzymol. *266*:554-571.

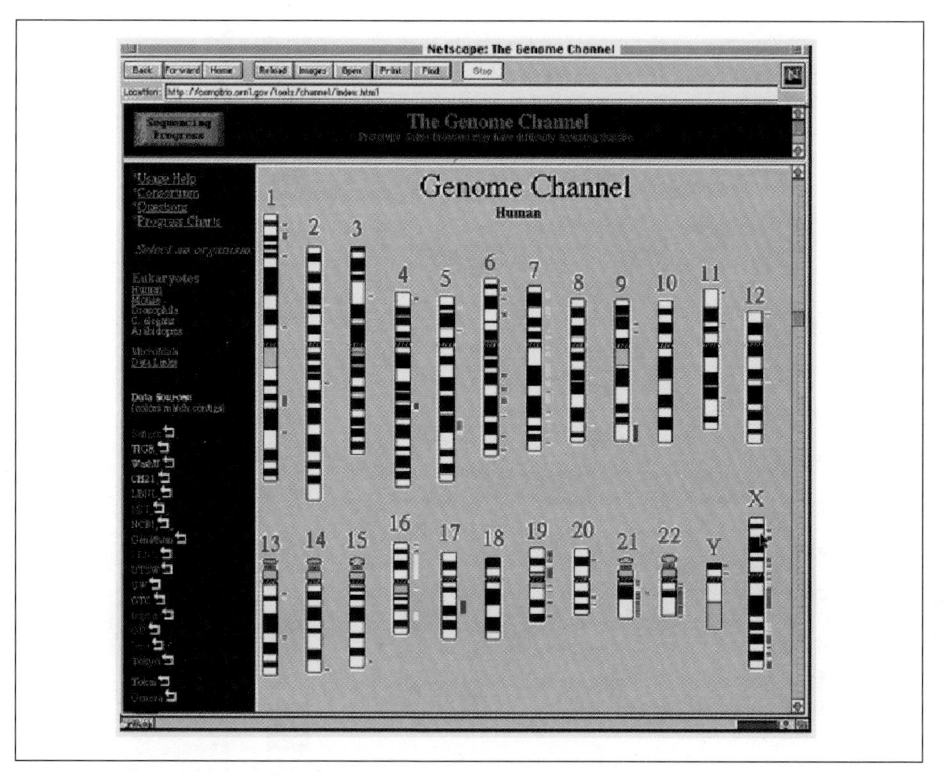

Chapter 2, Figure 1. The Genome Channel Web page maps progress on human genome sequencing.

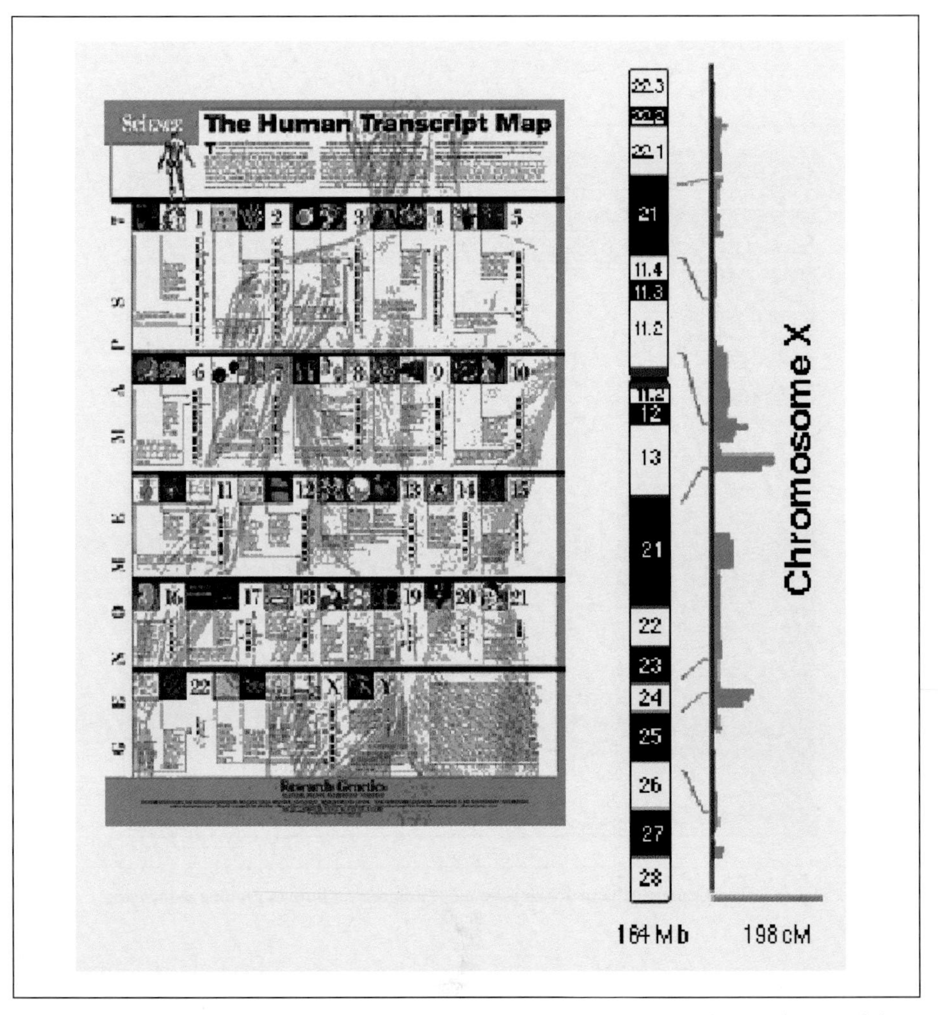

Chapter 2, Figure 2. NCBI Web page for the human genome transcript map and a sample map of the X chromosome.

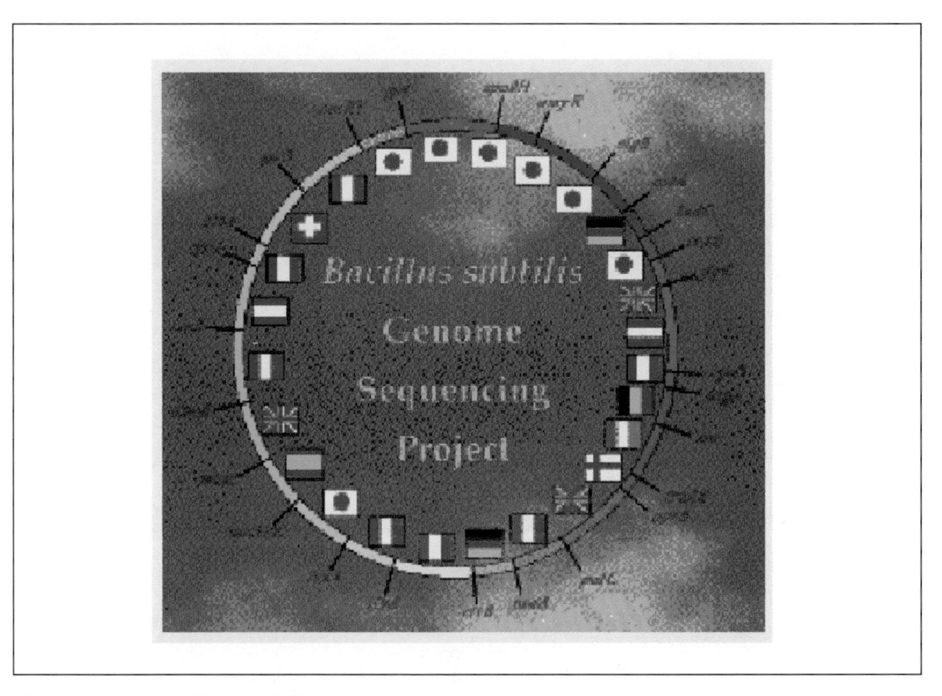

Chapter 2, Figure 3. The *B. subtilis* genome sequencing project.

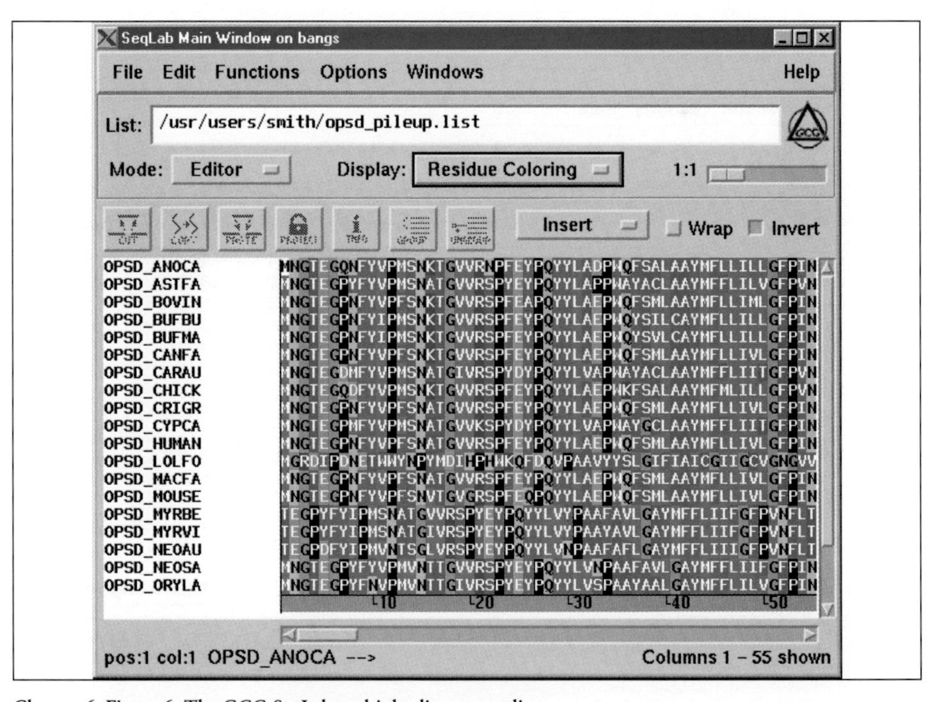

Chapter 6, Figure 6. The GCG SeqLab multiple alignment editor.

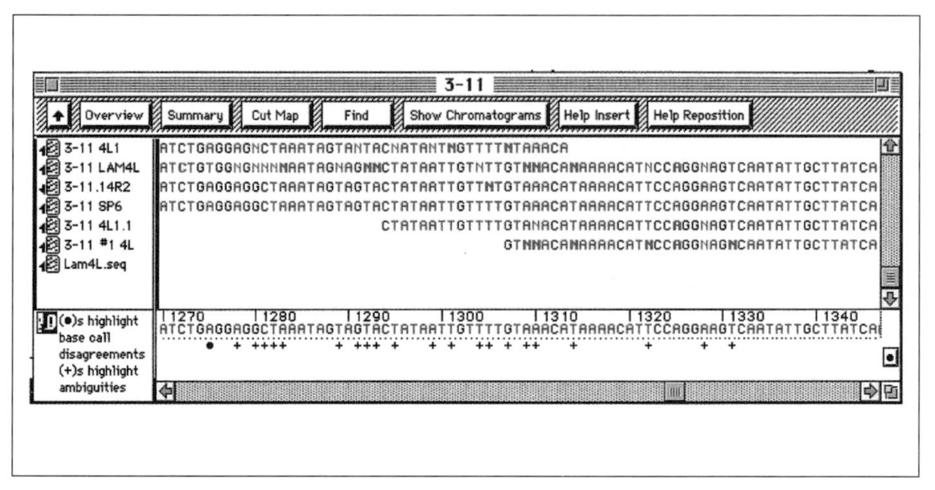

Chapter 6, Figure 7. Sequencher multiple alignment editor for DNA sequences.

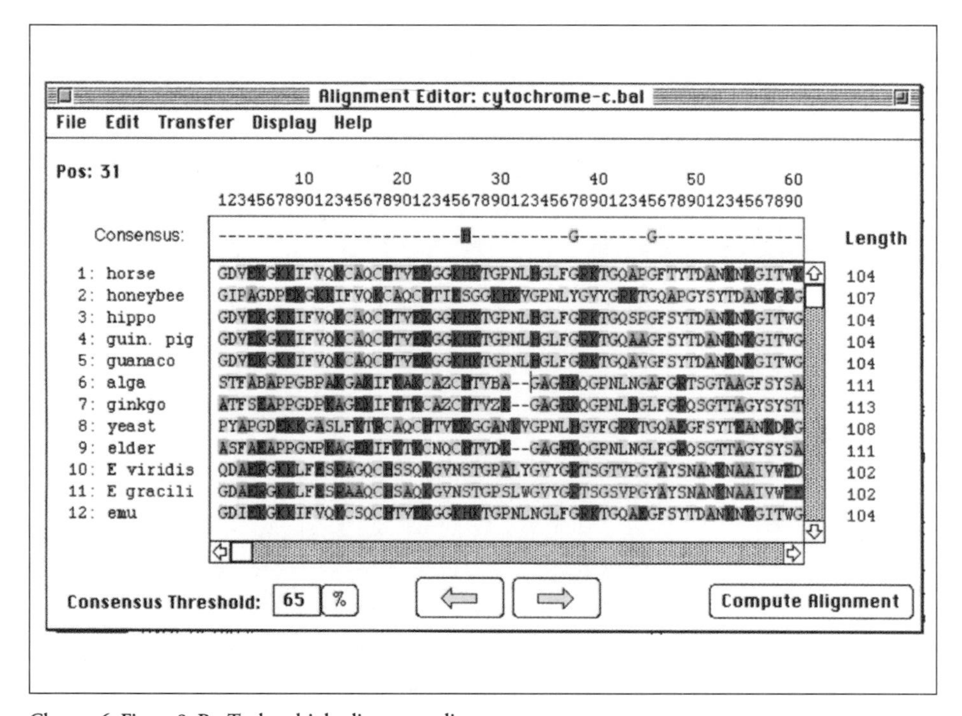

Chapter 6, Figure 8. PepTool multiple alignment editor.

Chapter 6, Figure 9. Sample alignment in CINEMA.

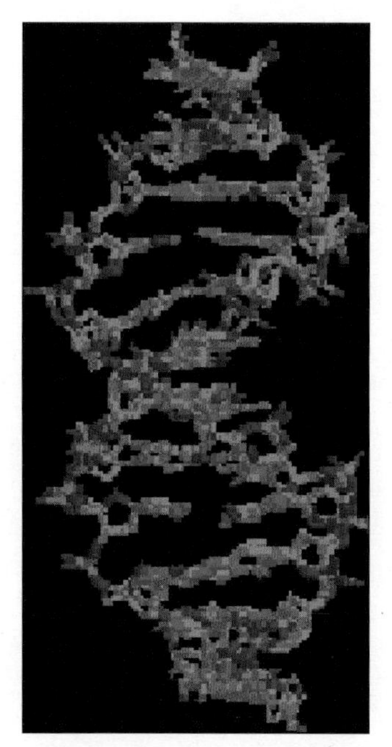

Chapter 7, Figure 1. DNA 3-dimensional structural model.

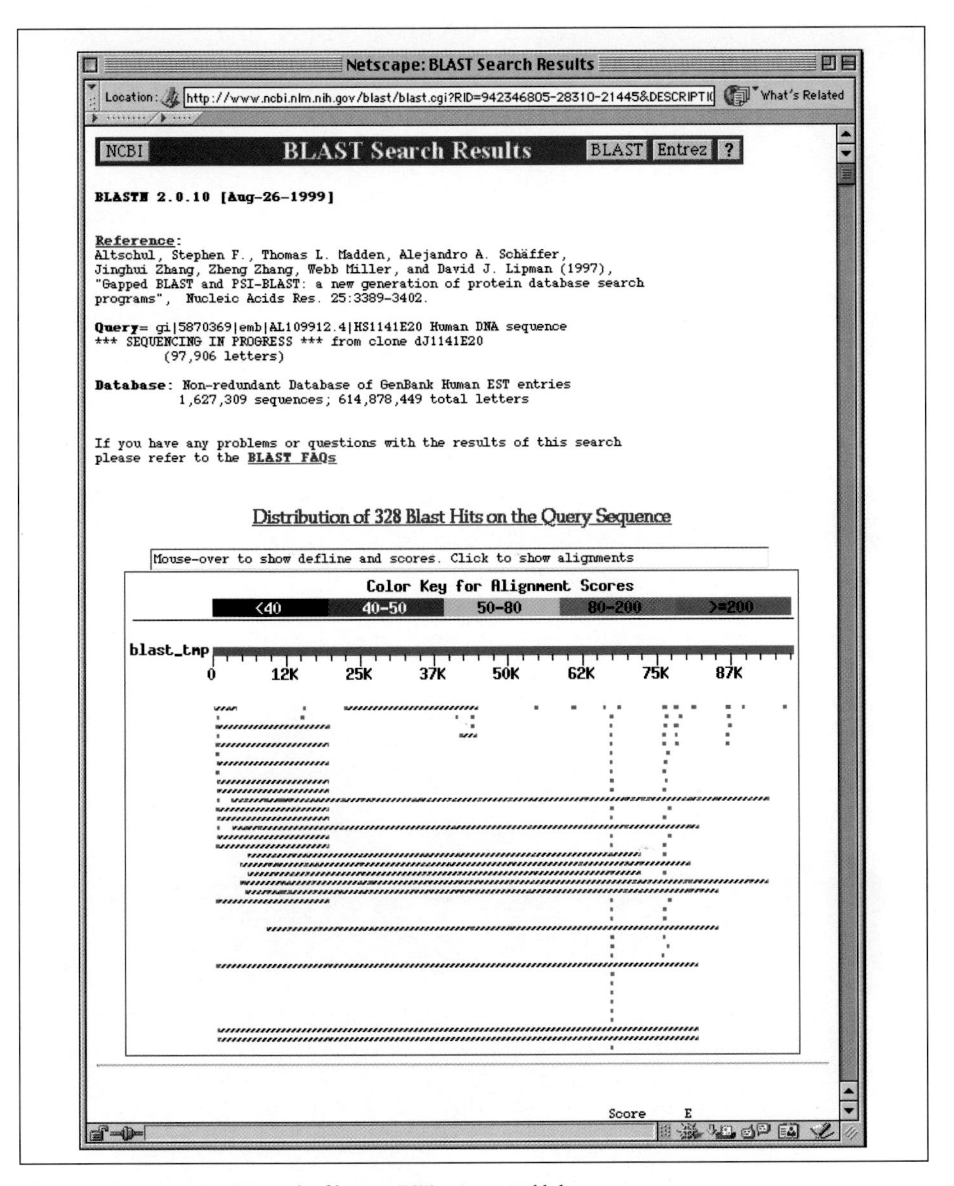

Chapter 7, Figure 2. BLAST search of human ESTs using a 97 kb human genome sequence.

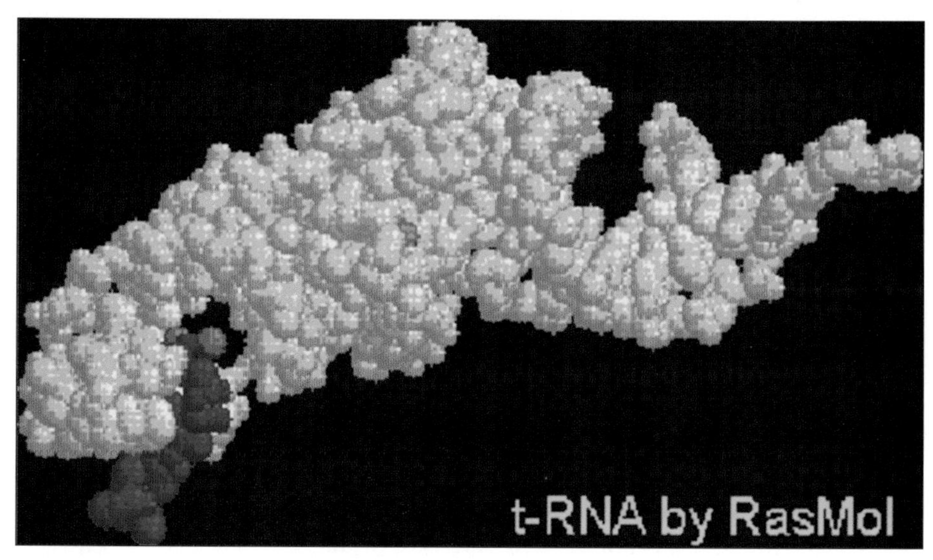

Chapter 7, Figure 3. Structural model of a tRNA molecule.

Chapter 7, Figure 4. Phage CRO repressor bound to DNA; courtesy of Andrew Coulson and Roger Sayles with RasMol, University of Edinburgh, 1993.

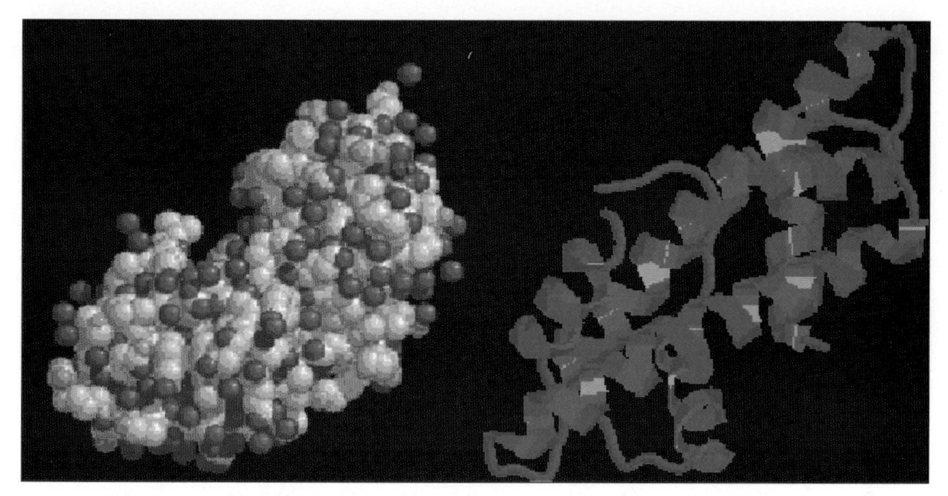

Chapter 7, Figure 5. Histone molecule shown by RasMol in space filling and ribbon models.

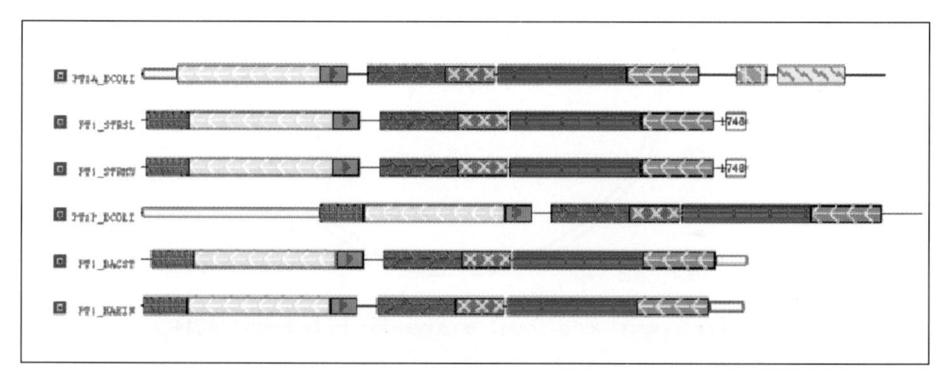

Chapter 7, Figure 9. Example of ProDom domains.

Links	Name	Strand	Expect
B S A ?	gb|U07604|LHU07604	+	1.4e-73
B S A ?	emb|X66723|LHDLDHYD	+	1.3e-72
B S A ?	emb|X60220|LDLDHA	+	4.6e-72
B S A ?	gb|M85228|LBALHDA	+	4e-71
B S A ?	gb|M26929|LBADDH	+	4.2e-62
B S A ?	dbj|D90339|LBADLDH	+	1.6e-61
B S A ?	gb|L29327|LEUDLDH	+	1e-56

Chapter 7, Figure 10. A sample MEME output.

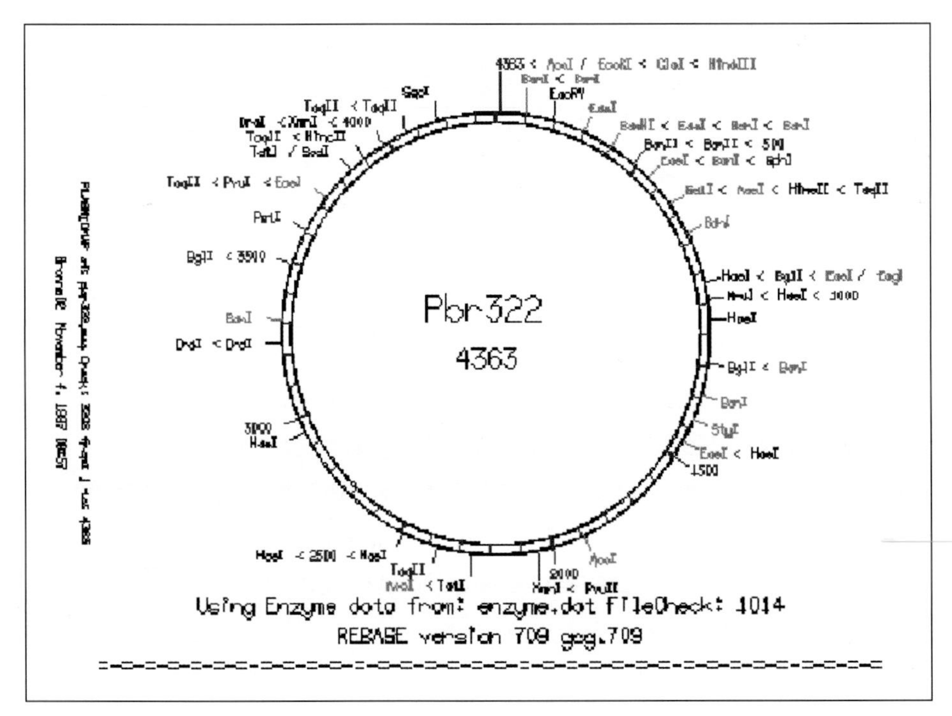

Chapter 8, Figure 2. A map of plasmid pBR322 created with the GCG program PLASMIDMAP.

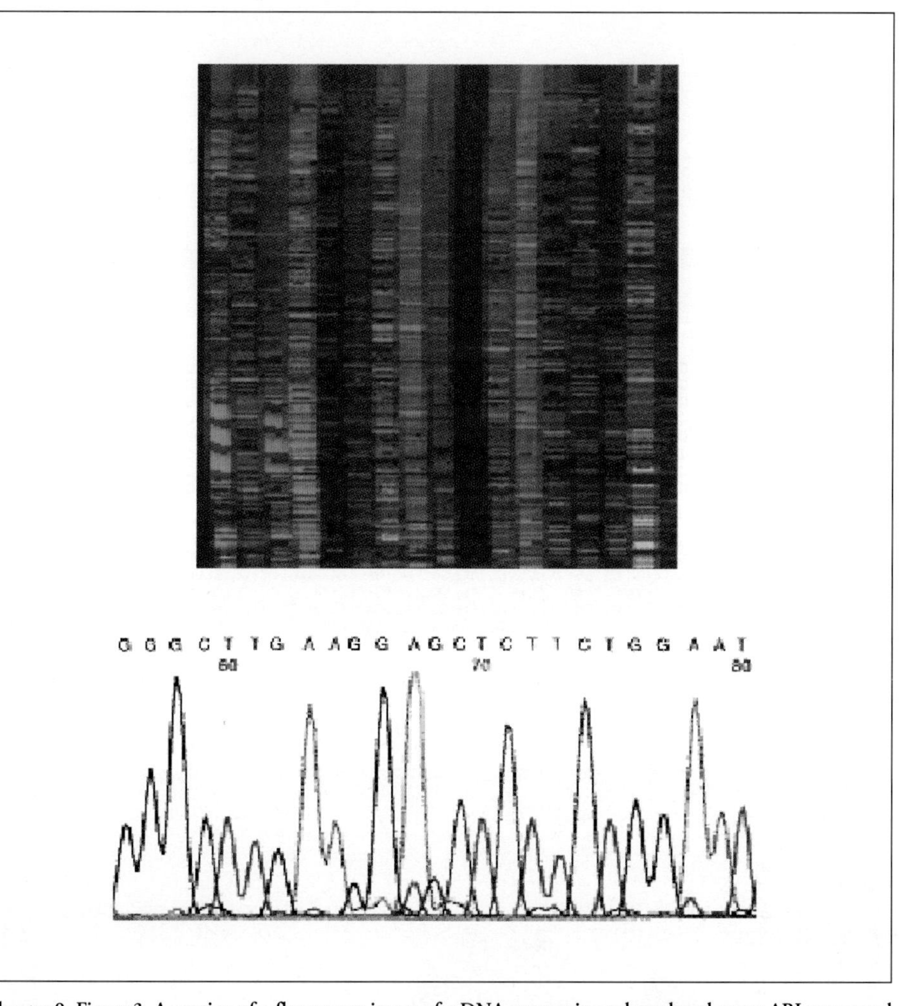

Chapter 9, Figure 3. A portion of a fluorescent image of a DNA sequencing gel produced on an ABI automated sequencer (upper panel) and a sample chromatogram from one lane of the gel (lower panel).

```
TGTGCTGGAATTCGGCTTGCTGCATCACCCCACTCAGCATCATCGTGCTCTGCTACCT
TGTCCTCATGGTC:ACCTGCTGCATCACCCCACTCAGCATCATCGTGCTCTGCTACCT
TGTCCTCATGGTC:ACCTGCTGCATCACCCCACTCAGCATCATCGTGCTCTGCTACCT
TGTCCTCATGGTC:ACCTGCTGCATCATCCCACTCGCTATCATCATGCTCTGCTACCT
TGTGCTGGAATTCGGCTTGCTGCATCACCCCACTCAGCATCATCGTGCTCTGTTACCT
450       |460      |470      |480      |490      |500
TGTCCTCATGGTC:ACCTGCTGCATCACCCCACTCAGCATCATCGTGCTCTGCTACCT
   •   •••••  •• •        •       •••      •       •
```

Chapter 9, Figure 7. A portion of a contig showing a consensus sequence and individual reads that differ from the consensus.

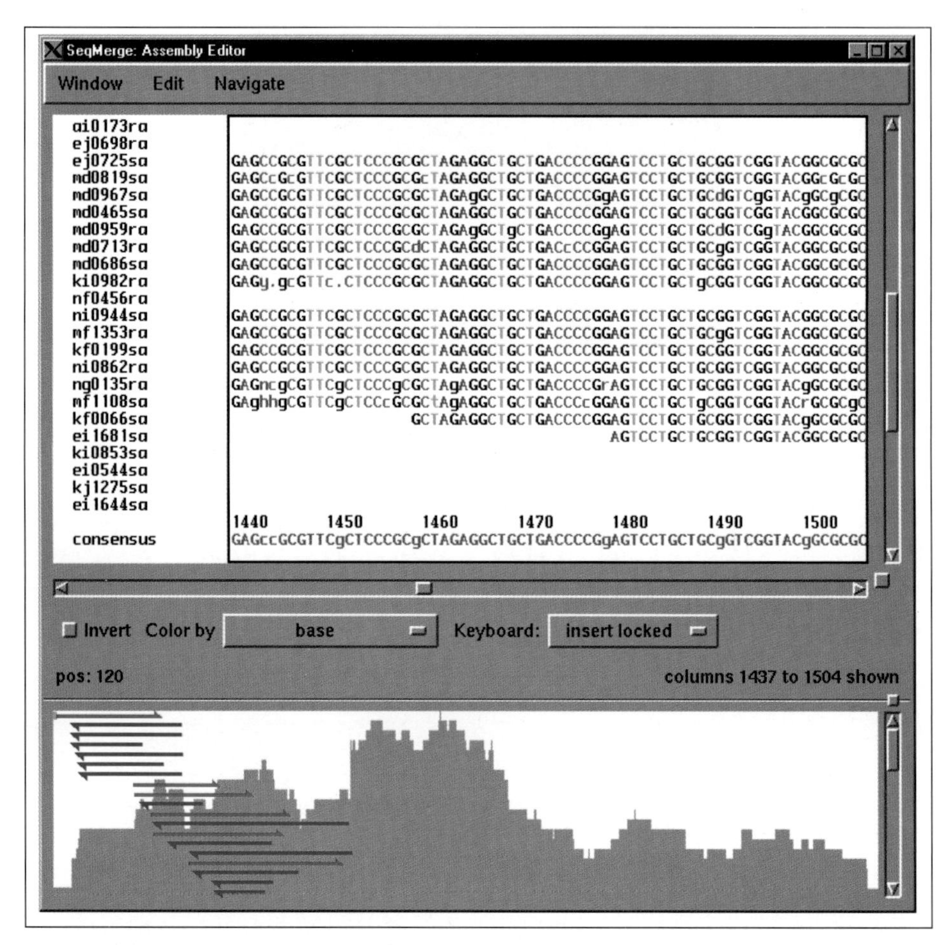

Chapter 9, Figure 8. Sample SeqMerge screen.

6 Pairwise and Multiple Alignment

INTRODUCTION TO ALIGNMENT

The alignment of two sequences (DNA or protein) is the fundamental building block of virtually all forms of sequence analysis including database searching, sequence assembly, and phylogenetics. Fortunately, from the standpoint of computational biology, this is a straightforward problem. Algorithms have been developed and computer programs written to rapidly and accurately align two sequences, and there is broad consensus among biologists that these programs do an excellent job.

In practice, making an alignment is very simple. Two sequences are written out, one on top of the other. Then, one sequence is moved with respect to the other, and gaps are inserted in each sequence to maximize identical pairs of bases (or similar pairs of amino acids) lining up on top of each other. The overall quality of the alignment is then evaluated based on a formula that counts the number of identical (or similar) pairs minus the number of mismatches and gaps. Gaps should be penalized significantly more than mismatches since an indiscriminate use of gaps can force an alignment between virtually any pair of sequences—leading to a meaningless result. However, the penalty assigned to a gap should not be proportional to the size of the gap because, from a biological perspective, larger insertion/deletions (indels) are almost as common as single-base indels. Alignment programs generally use separate penalty values for gap insertion (adding a new gap) and gap extension (increasing the size of a gap).

Alignment can be done by hand using any word processor or text editor to move lines of text and insert spaces, and this may be the best method when you wish to emphasize specific conserved domains (Figure 1). However, evaluating the quality and significance of an alignment is clearly a task for a computer program. In fact, alignment programs are able to simultaneously build an alignment and calculate the effect of each shift and gap insertion on its quality, then make decisions accordingly to produce an optimal alignment.

The optimal alignment produced by a computer program is not necessarily the best alignment from a biological perspective. The program is simply applying an

Bioinformatics
By Stuart M. Brown

algorithm and identifying the alignment that best fits the rules of that algorithm. It is also possible that many different (but similar) alignments may be almost equally good according to the scoring rules of a given algorithm. The results can be altered dramatically by changing the default values for gap insertion and extension penalties, or changing the matrix used to evaluate amino acid similarities. In fact, there is no single correct or "true" alignment for any pair (or group) of sequences. It is up to you, the biologist, to interpret and evaluate any alignment produced by a computer program and determine if it is useful. In general, the best alignment is one that reflects the underlying evolutionary similarity between the sequences. Since the actual evolutionary history of the sequences is unknown, and essentially unknowable, the evaluation of alignments is left to the intuition of biologists

Pairwise alignment is a relatively straightforward computational problem, and many different algorithms have been developed. The best solution seems to be an approach called dynamic programming (3). Dynamic programming breaks down a large problem into a series of smaller sub-problems, such that *(i)* the initial sub-problems are very simple; *(ii)* later problems can be calculated using the results of solutions found in the earlier stages; and *(iii)* the solution to the final sub-problem contains the overall solution.

The implementation of dynamic programming for pairwise sequence alignment can be described as the process of finding the best path through a simple dot plot comparison of two sequences (2). The two sequences are written on the x and y axes of a graph, and a path is drawn from the upper left corner of the graph to the lower right (Figure 2). Wherever the sequences are identical (or similar), the path moves diagonally; wherever the sequences differ, the path can move vertically or horizontally (inserting gaps in one sequence or the other), or it can allow a mismatch and move diagonally. At each step, the computer chooses the path that contains the alignment with the most identical (or similar) pairs and the fewest internal gaps. Thus each step is based on all of the previous steps that have brought the path to that particular spot. There may be more than one optimal path, and thus more than one optimal alignment, for a given pair of sequences.

```
CCTGTGAATCCTCTTGTG-AGACT
 ||| | || |||| ||| ||| | = 4 mismatches, 3 gaps, 17 identities
GCTG-GTATGCTCT-GTGGAGAGT

GC-TGGAATCTCTC----CGTGAG
|| |  |  | |||    || | | = 4 mismatches, 8 gaps, 12 identities
GCCT--AG-CTCTCACAACGCGTG
```

Figure 1. Sample alignment created with a text editor.

GLOBAL VERSUS LOCAL ALIGNMENTS

There are two different methods of applying dynamic programming to the alignment of two sequences. An optimal global alignment can be found for two complete sequences. This tends to average out the distribution of gaps throughout the entire length of the paired sequences and can mask smaller regions of high similarity. Global alignment algorithms are often not effective for highly diverged sequences and do not reflect the biological reality that two sequences may share only limited regions of conserved sequence. Sometimes two sequences may be derived from ancient recombination events where only a single functional domain is shared. The Genetics Computer Group (GCG) program GAP implements the Needleman and Wunsch (14) global alignment algorithm.

Alternatively, a local alignment can be found between the short region (or regions) of highest similarity between two sequences. The Smith and Waterman local alignment algorithm (22) is generally considered to be the best. This is implemented in the GCG program BESTFIT. There are a number of simple stand-alone pairwise alignment programs for various versions of the UNIX® operating system running on various types of computer hardware. In the FASTA sequence analysis package (available for most types of computers), global alignment by dynamic programming is implemented in the ALIGN program and local alignment in the LALIGN program (17).

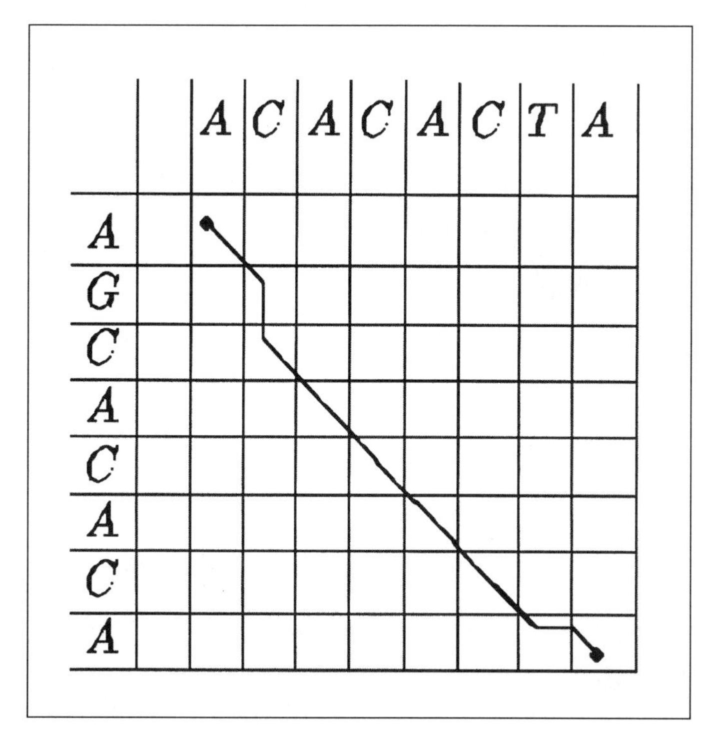

Figure 2. Graph of a dynamic programming decision matrix.

Global and local alignments will often give very different results for the alignment of the same two sequences. Sometimes the best approach to aligning two sequences requires a combination of local and global algorithms, particularly when the two sequences are of different lengths, or when an alignment is desired for the entire region surrounding a conserved domain. The local alignment will find the region of greatest similarity between the two sequences. Then, both sequences can be edited down to a range of bases (or amino acids) of the same length that surrounds the most similar region. These shorter sequence fragments can then be aligned with a global algorithm (Figure 3).

A variety of Web sites provide free pairwise alignment services (both local and global).

University of Southern California; Pairwise alignment server (both global and local algorithms).
(http://www-hto.usc.edu/software/seqaln/seqaln-query.html)

Baylor College of Medicine, Houston, TX; BCM Search Launcher (20).
(http://dot.imgen.bcm.tmc.edu:9331/seq-search/alignment.html)

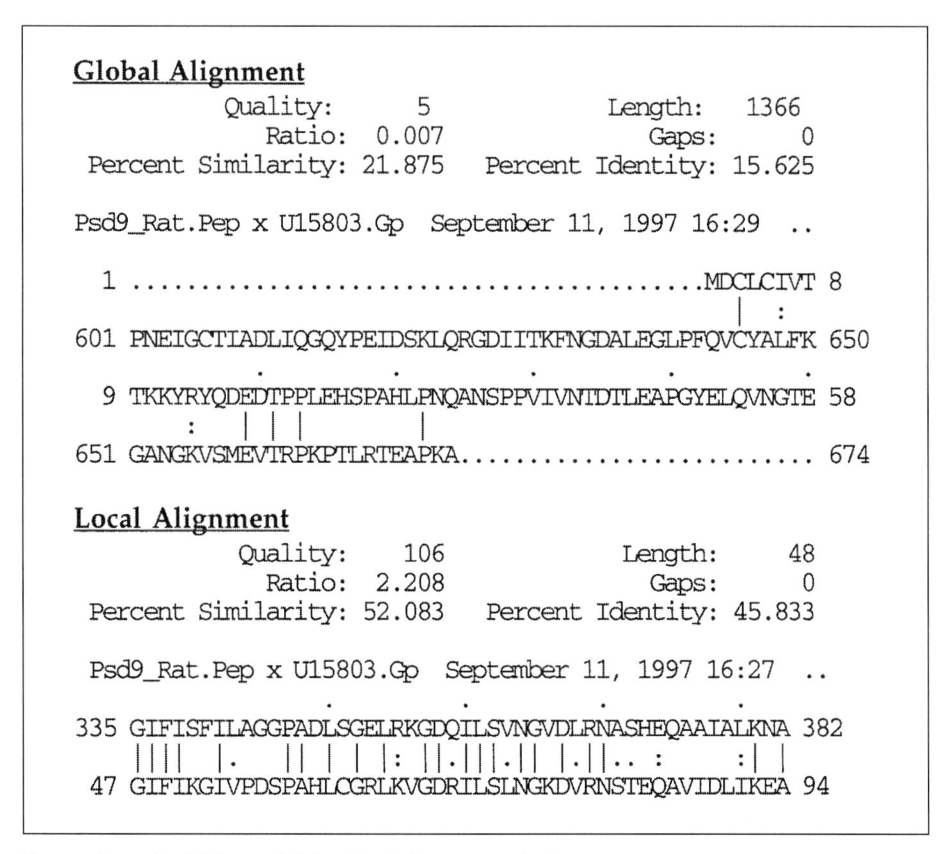

Figure 3. Example of different global and local alignment results for two sequences.

Ecole pour les Etudes et la Recherche en Informatique et Electronique (EERIE), Nimes, France; GeneStream server.
(http://genome.eerie.fr/)

Centre de Recherche en Biochimie Macromoléculaire (CRBM), Montpellier, France; GeneStream server.
(http://vega.crbm.cnrs-mop.fr/)

University of Southampton, England, UK; Southampton Bioinformatics Data Server (SBDS).
(http://molbiol.soton.ac.uk/compute/align.html)

MULTIPLE SEQUENCE ALIGNMENT

In theory, making a true optimal alignment among many large sequences is computationally almost impossible. Imagine that an alignment of two sequences defines a dot plot type graph, then three sequences would define a cube, and each additional sequence would end another dimension, so that the complexity of the problem increases exponentially with the number of sequences involved. The dynamic programming approach of the Smith-Waterman (22) and Needleman-Wunsch (14) methods still apply in multiple dimensions; however, the problem is computationally very demanding. That is, while one could in theory do an N-dimensional Smith-Waterman analysis, in practice, except on the simplest cases, it would take more computer power than is generally available.

Rather than computing the absolute optimal alignment of multiple sequences from all of the nearly infinite possibilities, it is possible to estimate the best alignment using a progressive pairwise approach (5,9). This is the method used by almost all of the current multiple alignment programs including the GCG program PILEUP, as well as the free stand-alone program CLUSTALW (8,24), and several programs available over the Web including PIMA (19), MSA (13), MAP (11), and ToPLign (23). The disadvantage of the progressive pairwise approach is that the alignment that is produced is only an estimate and is not guaranteed to be the optimal alignment.

All of the progressive pairwise methods use the same strategy, initially developed by Feng and Doolittle (5). First, rough similarity scores are calculated between all pairs of sequences to be aligned using a fast pairwise alignment algorithm, and they are clustered into a dendrogram structure. Then, the most similar pairs of sequences are aligned using either a global (PILEUP) or a local (CLUSTALW, PIMA) pairwise dynamic programming method. This aligned pair is then treated as a single sequence (a consensus), which is aligned with the next most similar sequence. The final multiple alignment is performed by a series of pairwise alignments between sequences and clusters of sequences, according to the branching order in the dendrogram. The final alignment will usually change somewhat if a different comparison matrix is used, or if the gap penalties are changed.

Before you run any alignment program, it is necessary to study the sequences that will be aligned. If a set of sequences are of different lengths, gaps will be added to the ends of all shorter sequences to make them equal to the longest one in the set. Many multiple alignment programs have an optional setting that can penalize gaps at the ends of sequences either the same way as internal gaps or give no penalty to end gaps. If you are aligning a bunch of different proteins, and you know some regions of some proteins are just not at all similar to the rest, cut those regions out before you do the alignment, especially if the dissimilar regions are on the ends of the proteins. Similarly, if you are interested just in some particular repeat or motif, extract it from the original sequence as best you can and then do the alignment. The reason is, everything you throw into an alignment that does not belong there just acts to gunk up the works. The final alignment may still come out right, but then, it might not.

The final output produced by PILEUP or CLUSTAL is a text file in multiple sequence format (MSF). It is simply a list of one sequence on top of another with gaps inserted in each sequence where appropriate (Figure 4). This MSF file can be used as input for a variety of programs that work with aligned sequences, such as phylogenetics and pattern analysis programs, or it can be transferred to a desktop computer for editing and formatting.

PILEUP has an option to output a figure of the similarity tree (dendrogram) that it used to guide the alignment process. It is usually a good idea to look at it, just to make sure that the order of alignment makes some sort of sense—it can help you catch misnamed sequences, for instance. That tree may look like a phylogenetic tree, but it is not; it is simply a clustering of sequences based on the

```
//
               1                                                     50
Pa29_Pseau    NLIQFKSIIE CANRGSRRWL DYADYGCYCG WGGSGTPVDE LDRCCKVHDE
Pa2b_Psepo    NLIQFSNMIK CAIPGSRPLF QYADYGCYCG PGGHGTPVDE LDRCCKIHDD
 Pa2_Aipla    NLYQFDNMIQ CANKGKRATW HYMDYGCYCG SGGSGTPVDA LDRCCKAHDD
Pa2x_Notsc    NVAQFDNMIE CANYGSRPSW HYMEYGCYCG KEGSGTPVDE LDRCCKAHDD
Pa21_Oxysc    NLLQFGFMIR CANRRSRPVW HYMDYGCYCG KGGSGTPVDD LDRCCQVHDE
Pa2a_Psete    NLVQFSYLIR CANKYKRPGW HYANYGCYCG SGGRGTPVDD VDRCCQAHDK
Pa2b_Psete    DLVEFGFMIR CANRNSQPAW QYMDYGCYCG KRGSGTPVDD VDRCCQTHNE
Pa24_Mouse    ~~~~FQRMVK .HVTGRSAFF SYYGYGCYCG LGGKGLPVDA TDRCCWAHDC
  Pa24_Rat    ~~~~FQRMVK .HITGRSAFF SYYGYGCYCG LGGRGIPVDA TDRCCWAHDC
Pa22_Bitna    DLTQFGNMIN .KMG..QSVF DYIYYGCYCG WGGQGKPRDA TDRCCFVHDC
 Pa2_Bitga    DLTQFGNMIN .KMG..QSVF DYIYYGCYCG WGGKGKPIDA TDRCCFVHDC
Pa2b_Trifl    SLVQLWKMIF .QETGKEAAK NYGLYGCNCG VGRRGKPKDA TDSCCYVHKC
Pa2b_Trimu    SLIELGKMIF .QETGKNPVK NYGLYLCNCG VGNRGKPVDA TDRCCFVHKC
Pa2m_Agkcl    SLLELGKMIL .QETGKNAIT SYGSYGCNCG WGHRGQPKDA TDRCCFVHKC
Pa2h_Agkpi    SVLELGKMIL .QETGKNAIT SYGSYGCNCG WGHRGQPKDA TDRCCFVHKC
Pa22_Botas    SLFELGKMIL .QETGKNPAK SYGAYGCNCG VLGRGKPKDA TDRCCYVHKC
```

Figure 4. MSF file format.

number of pairwise differences. A rigorous phylogenetic analysis of the aligned sequences will often produce a phylogenetic tree that is similar to the dendrogram created by PILEUP (Figure 5), but this is not always the case. CLUSTAL also has the option to print out a dendrogram of sequences, but this is also a clustering based on a simple distance measure, not a true phylogenetic tree.

GCG has developed an X Windows-based interface called SeqLab, based on Steve Smith's Genetic Data Environment (21), which works extremely well for multiple sequence alignment projects, coupling the power of a mainframe with the graphical interface of an X-terminal (or a Macintosh®/Windows® computer with an X-Windows emulator). Lists of sequences produced by similarity search programs such as FASTA can be directly imported into the SeqLab multiple sequence editor. Then, any set of sequences or regions within sequences can be aligned with PILEUP. The resulting multiple sequence alignments can be loaded into the sequence editor (Figure 6), and the alignment can be refined by inserting and deleting gaps. The only weak part of the program is its limited ability to format and print out the alignments.

GCG's PILEUP program uses a global alignment method, so it is does not do a good job of aligning sequences that share a region of similarity (a conserved domain) but are not similar throughout their whole length. PILEUP is also limited in the total number of gaps allowed in a multiple sequence alignment, 2000 by default. If one long sequence is included in an alignment with many shorter ones, end gaps will be assigned to each short sequence, and the program will fail

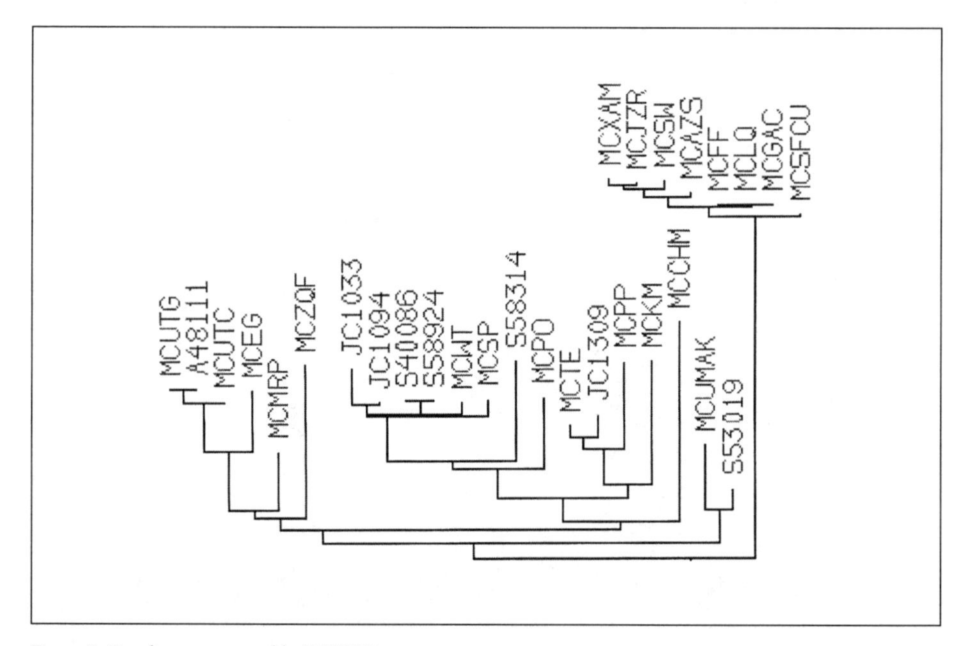

Figure 5. Dendrogram created by PILEUP.

to produce a result. However, by designating appropriate start and end sites for each of the sequences to be aligned, and with the appropriate use of comparison matrix and gap penalties, PILEUP does at least as well as any of the other programs, aligning sequences that are homologous throughout their entire length.

There are several useful and fairly powerful alignment programs for Macintosh and Windows computers. The best free programs are MACAW (18) and CLUSTAL (24). MACAW (Multiple Alignment Construction and Analysis Workbench) is slow in computing alignments of more than a handful of sequences of any significant length, but it has very useful tools for formatting and in-depth study of short blocks of aligned sequences. It can be used to examine and refine multiple alignments created by more powerful computers.

CLUSTAL is the most flexible and most easily available of the multiple alignment programs. There is both a command line interface (CLUSTALW) and a graphical interface (CLUSTALX) available for almost all computer platforms including Macintosh, Windows® 3.x, Windows 95®, Windows NT, UNIX, and VMS®. CLUSTAL is also available over the Web from many different servers (see below).

CLUSTAL is an excellent general purpose multiple alignment program for DNA or proteins. It can align multiple sequences to produce both multiply aligned

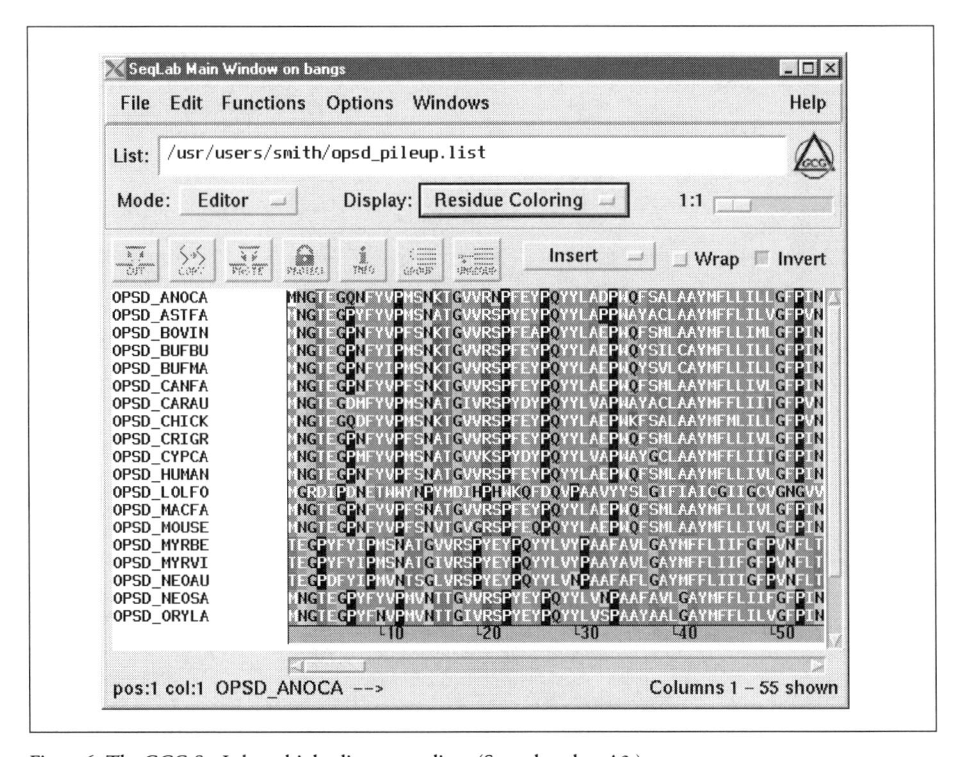

Figure 6. The GCG SeqLab multiple alignment editor. (See color plate A3.)

text files and phylogenetic trees. Sequences can be imported in the following file formats: National Biomedical Research Foundation (NBRF)/Protein Information Resource (PIR), European Molecular Biology Laboratory (EMBL)/SWISS-PROT, Pearson (FASTA), CLUSTAL (*.aln), GCG/MSF (PILEUP), and Genetic Data Environment (GDE) flat file. CLUSTAL can align groups of already aligned sequences to each other, or add new sequences to existing alignments. It is also possible to re-align selected sequences or selected regions of the alignment leaving the unselected portions of the alignment constant. Secondary structures can be used to guide the alignments. Alignments can be output in color as postscript files.

CLUSTAL allows the user to control both the methods and the parameters (distance matrixes and gap penalties) used in making multiple alignments. The calculation of the initial pairwise distances between all pairs of sequences (used to build the phylogenetic tree that guides the final multiple alignment) can be done using either the slower dynamic programming method (Smith-Waterman) or by the rapid method of Wilbur and Lipman (FASTA) (25). Gap penalties in the final multiple alignment can be adjusted based on the amino acid composition of specific regions of a sequence, for example, reduced penalties in hydrophilic regions.

Several of the full-featured commercial sequence analysis packages for PCs also have multiple alignment functions. MacVector™ (Oxford Molecular Group, Oxford, England, UK; Macintosh only) can perform alignments of up to 30 DNA or protein sequences using the CLUSTALW algorithm. The output of the multiple sequence alignment can be saved as PICT files for use in publications. OMIGA (Oxford Molecular, Windows 95/NT only) also has an integrated version of CLUSTAL. Sequencher (Gene Codes, Ann Arbor, MI, USA; Macintosh and Windows 95/NT) has a superb multiple alignment editor, but its alignment algorithm is designed for assembling contigs from DNA sequencing reads (Figure 7). It cannot align protein sequences nor can it align highly diverged sequences.

The PepTool program, created by BioTools (Edmonton, Alberta, Canada), also includes tools for creating and editing multiple alignments. The alignment algorithm is a progressive pairwise method called XALIGN (26), which is very similar to PILEUP (Figure 8). The alignment editor provides many useful functions including highlighting of columns of identical amino acids (with an adjustable cutoff for percent identity), displaying a consensus sequence, and coloring amino acid residues by their chemical properties. Gaps can be inserted anywhere in the aligned sequences, and entire blocks of sequences can be shifted. Individual sequences can be added to existing alignments, and alignments can be annotated with structural information (added by the user to the individual sequence files before alignment).

There are many Web servers that offer multiple sequence alignment programs, but they each have limitations.

- Baylor College of Medicine, Houston, TX, USA offers CLUSTALW, MAP, and PIMA at its BCM Search Launcher Web site. Limitations are 20 000 characters and 60 minutes of computing time.

(http://dot.imgen.bcm.tmc.edu:9331/multi-align/multi-align.html)

- CLUSTALW is available over the Web at the European Bioinformatics Institute, Cambridge, England, UK.
(http://www2.ebi.ac.uk/clustalw)

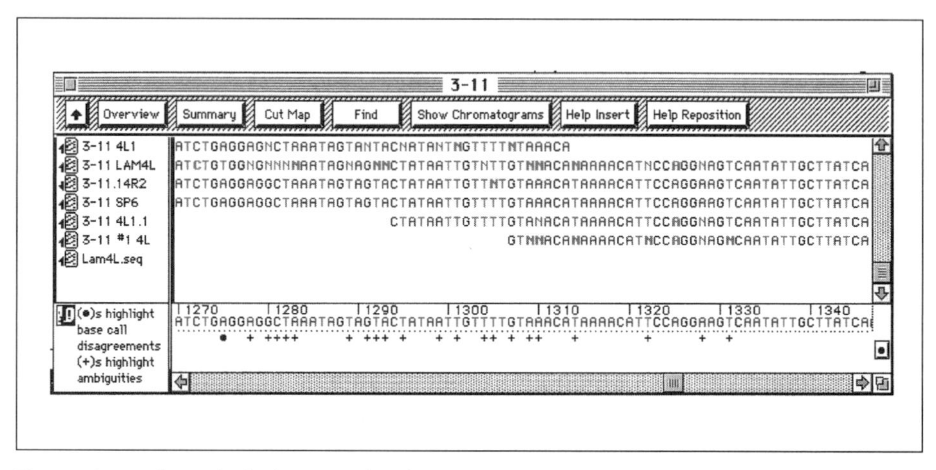

Figure 7. Sequencher multiple alignment editor for DNA sequences. (See color plate A4.)

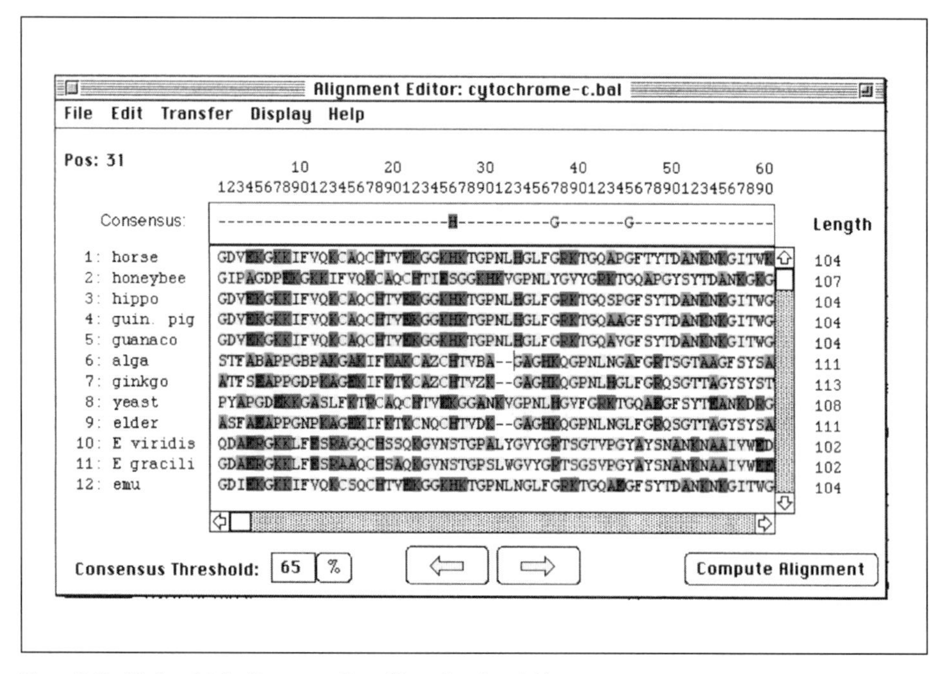

Figure 8. PepTool multiple alignment editor. (See color plate A4.)

- Washington University, St. Louis, MO, USA, offers a Web server running CLUSTALW, MSA, and ctree.
  (http://www.ibc.wustl.edu/msa)
- The ETHZ server of the Swiss Federal Institute of Technology, Zurich, Switzerland, offers a multiple alignment function called ALLALL, which uses a progressive pairwise implementation of the dynamic programming algorithm applied to a concept called probabilistic ancestral sequence:
  (http://cbrg.inf.ethz.ch/subsection3_1_1.html)
- The Multiple Alignment Resource Page at the VSNS BioComputing Division, hosted at the University of Bielefeld, Germany, has links to many others.
  (http://www.techfak.uni-bielefeld.de/bcd/Curric/MulAli/welcome.html)

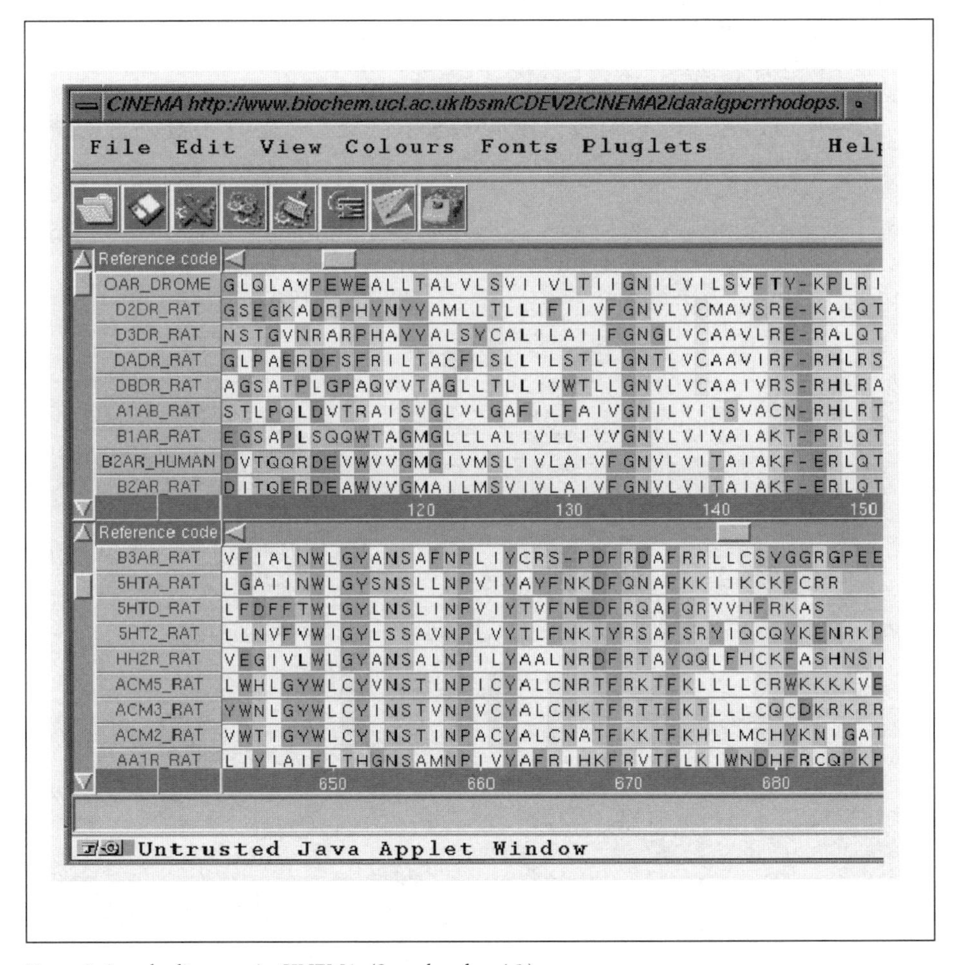

Figure 9. Sample alignment in CINEMA. (See color plate A5.)

EDITING AND FORMATTING MULTIPLE ALIGNMENTS

Multiple sequence alignment is a mathematically complex process. For a given group of sequences, there is no single correct alignment, only an alignment that is optimal according to some set of measurements. The multiple alignment programs such as PILEUP and CLUSTAL rely on approximations, so it is not even possible to be sure that the alignments they produce are truly optimal. The bottom line is that determining which alignment is best for a given set of sequences is really up to the judgment of the investigator. You should feel free to tweak the parameters used by the alignment programs (gap penalties, protein scoring matrixes, etc.), but also to adjust the alignment itself to conform to your ideas about the relationships among the sequences.

There are a variety of multiple sequence editors that can be used to modify a multiple alignment. These programs can be very useful in formatting and annotating an alignment for publication. An editor can also be very useful to make some modifications by hand to improve biologically significant regions in a multiple alignment created by one of the automated alignment programs.

Multiple alignments created in GCG with PILEUP can be edited with LINEUP, and consensus sequences can be created with PRETTY and PRETTYBOX. The MACAW (18) and SeqVu (6) programs for Macintosh, the GeneDoc program (15) for Windows PCs, and *DCSE* (4) for DOS and UNIX are free and provide excellent editor functionality. It is often best to use one of these Mac or PC programs to put the final tweaks on an alignment to produce a publication-

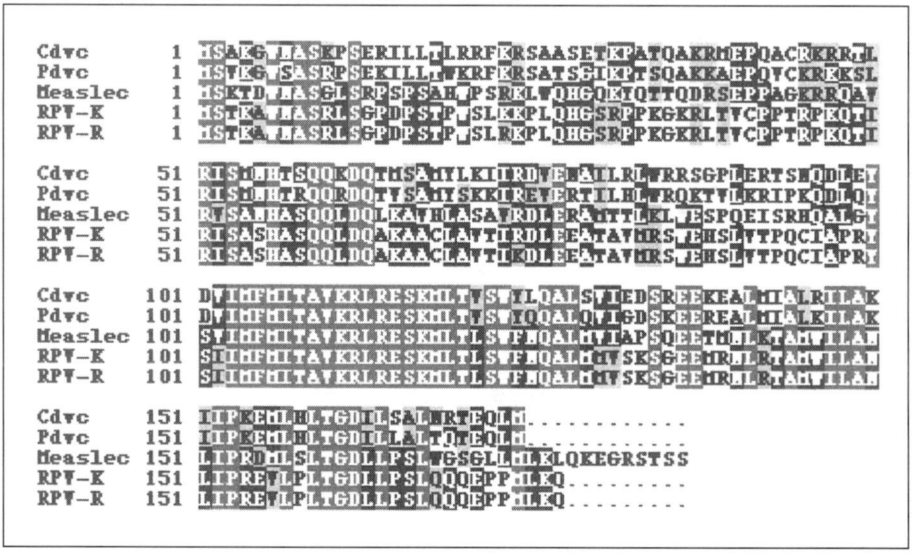

Figure 10. Shaded alignment produced by MacBox (similar to the output from BOXSHADE and PRETTYBOX).

quality figure, particularly if you want to use color to emphasize relationships among related amino acids.

There is a Web-based program for editing multiple alignments called CINEMA (color interactive editor for multiple alignments), which has been created completely in the Java™ computer language (16). The CINEMA program can be launched as a self-contained applet, which runs as a program on your desktop computer (not on the Web server) (Figure 9). DNA and protein alignments are edited by clicking on the sequences and dragging them to create gaps; whole sequences may be shifted to the left or right by clicking on the right mouse button and dragging. Amino acid sequences are colored according to residue type (acidic, basic, aromatic, etc.). It is also possible to change the colors.

(**http://www.biochem.ucl.ac.uk/bsm/dbbrowser/CINEMA2.1/**)

Once a multiple alignment is created and tweaked to your satisfaction, there is still the problem of producing a publication-quality printout. Investigators frequently wish to add special formatting to alignments such as blocks of shading to indicate regions of homology, bold and underlined regions of text to indicate primers, functional regions, etc. There is no single perfect solution to this problem.

On mainframe computers (UNIX and VMS systems), there is a stand-alone program called BOXSHADE (10) that can be used to apply a variety of types of shading to the MSF files produced by PILEUP and CLUSTAL. Another similar program called ShadyBox (12) is available for UNIX systems to produce shaded postscript output files. GCG has a program called PRETTY that can be used to calculate consensus sequences, change the letters in the sequences within an alignment between upper and lowercase, depending on whether they match the consensus, or replace all letters that are identical to the consensus with dots, leaving only the differences. There is another GCG program called PRETTYBOX that combines the features of PRETTY with many of the features provided by BOXSHADE, but produces only postscript output files. MacBox (1) is a version of the BOXSHADE program for the Macintosh (Figure 10).

REFERENCES

1. **Baron, M.D.** 1998. MacBoxshade: a Macintosh implementation of BOXSHADE. Freely distributed by the author for educational and noncommercial research purposes: (**ftp://www.isrec.isb-sib.ch/pub/sib-isrec/boxshade**).
2. **Barton, G.J.** 1996. Protein sequence alignment and database scanning. *In* M.J.E. Sternberg (Ed.), Protein Structure Prediction: A Practical Approach. IRL Press at Oxford University Press, Oxford.
3. **Bellman, R., J. Holland, and R. Kalaba.** 1959. On an application of dynamic programming to the synthesis of logical systems. J. ACM *6*:486-493.
4. **De Rijk, P. and R. De Wachter.** 1993. DCSE v2.54, an interactive tool for sequence alignment and secondary structure research. Comput. Appl. Biosci. *9*:735-740.
5. **Feng, D.F. and R.F. Doolittle.** 1987. Progressive sequence alignment as a prerequisite to correct phylogenetic trees. J. Mol. Evol. *25*:351-360.
6. **Gardner, J.** 1996. SeqVu. Distributed by the author as shareware. The Garvan Institute of Medical Research, Sydney, Australia.
7. **Gonnet, G.H.** 1994. A tutorial introduction to computational biochemistry using Darwin. E.T.H. Zurich, Switzerland.

8. **Higgins, D.G. and P.M. Sharp.** 1988. CLUSTAL: a package for performing multiple sequence alignment on a microcomputer. Gene *73*:237-244.
9. **Higgins, D.G. and P.M. Sharp.** 1989. Fast and sensitive multiple sequence alignments on a microcomputer. Comput. Appl. Biosci. *5*:151-153.
10. **Hofmann, K. and M.D. Baron.** 1999. BOXSHADE 3.3.3: pretty printing and shading of multiple-alignment files. Freely distributed by the authors for educational and noncommercial research purposes: (**ftp://www.isrec.isb-sib.ch/pub/sib-isrec/boxshade**).
11. **Huang, X.** 1994. On global sequence alignment. Comput. Appl. Biosci. *10*:227-235.
12. **Huynh, C.** 1995. ShadyBox. Australian National Genomic Information Service. Freely distributed by the author for educational and noncommercial research purposes: (**ftp://morgan.angis.su.oz.au/pub/unix**).
13. **Lipman, D.J., S.F. Altschul, and J. Kececioglu.** 1989. A tool for multiple sequence alignment. Proc. Natl. Acad. Sci. USA *86*:4412-4415.
14. **Needelman, S.B. and C.D. Wunch.** 1970. A general method applicable to the search for similarities in the amino acid sequence of two proteins. J. Mol. Biol. *48*:443-453.
15. **Nicholas, K.B., H.B. Nicholas, and D.W. Deerfield.** 1997. GeneDoc: analysis and visualization of genetic variation. EMBNEW.NEWS *4*:14. Freely distributed by the author for educational and noncommercial research purposes: (**http://www.cris.com/~Ketchup/genedoc.shtml**).
16. **Parry-Smith, D.J., A.W. Payne, A.D. Michie, and T.K. Attwood.** 1998. CINEMA—a novel colour INteractive editor for multiple alignments. Gene *221*:GC57-63.
17. **Pearson, W.R. and D.J. Lipman.** 1988. Improved tools for biological sequence comparison. Proc. Natl. Acad. Sci. USA *85*:2444-2448.
18. **Schuler, G.D., S.F. Altschul, and D.J. Lipman.** 1991. A workbench for multiple alignment construction and analysis. Proteins *9*:180-190.
19. **Smith, R.F. and T.F. Smith.** 1992. Pattern-induced multi-sequence alignment (PIMA) algorithm employing secondary structure-dependent gap penalties for use in comparative protein modelling. Protein Eng. *5*:35-41.
20. **Smith, R.F., B.A. Wiese, M.K. Wojzynski, D.B. Davison, and K.C. Worley.** 1996. BCM Search Launcher—an integrated interface to molecular biology data base search and analysis services available on the World Wide Web. Genome Res. *6*:454-462.
21. **Smith, S.W., R. Overbeek, C.R. Woese, W. Gilbert, and P. Gillever.** 1994. The genetic data environment: an expandable GUI for multiple sequence analysis. Comput. Appl. Biosci. *10*:671-675.
22. **Smith, T.F. and M.S. Waterman.** 1981. Identification of common molecular subsequences. J. Mol. Biol. *147*:195-197.
23. **Thiele, R., R. Zimmer, and T. Lengauer.** 1995. Recursive dynamic programming for adaptive sequence and structure alignment, p. 384-392. *In* C. Rawlings et al. (Eds.). Proc. of the Third International Conference on Intelligent Systems for Molecular Biology. AAAI Press, Menlo Park, CA.
24. **Thompson, J.D., D.G. Higgins, and T.J. Gibson.** 1994. CLUSTALW: improving the sensitivity of progressive multiple sequence alignment through sequence weighting, positions-specific gap penalties and weight matrix choice. Nucleic Acids Res. *22*:4673-4680.
25. **Wilbur, W.J. and D.J. Lipman.** 1983. Rapid similarity searches of nucleic acid and protein data banks. Proc. Natl. Acad. Sci. USA *80*:726-730.
26. **Wishart, D.S., R.F. Boyko, and B.D. Sykes.** 1994. Constrained multiple sequence alignment using XALIGN. Comput. Appl. Biosci. *10*:687-688.

7 Structure–Function Relationships

PROTEIN STRUCTURE AND FUNCTION

The Genome Projects are very rapidly generating a huge amount of DNA sequence data for many different organisms. However, this raw DNA data is really only useful to biologists after it is processed, annotated, and interpreted. Once a big chunk of genomic DNA sequence is obtained, the first question of interest to a biologist is: "Where are the genes?" Next would be: "What proteins do these genes encode?" And then comes the real kicker: "What are the functions of these proteins?"

DNA sequences have been called the blueprints for life. This is an accurate metaphor, but the blueprints are in code. The Central Dogma of molecular biology as theorized by Watson and Crick in their double helix model (Figure 1) (57) is really a set of instructions for reading this code: DNA is transcribed into RNA, and RNA is translated into protein. The proteins then perform metabolic, structural, and regulatory functions in the cell. Cellular biochemistry works using the language of interactions between 3-D molecular structures, so it is the 3-D structure of a protein that determines its function. Therefore, the relationship of sequence to function is primarily concerned with understanding the 3-D folding of proteins and inferring protein function from these 3-D structures. The decoding of the simple amino acid sequences of proteins into the language of protein structure and function is the most difficult challenge of bioinformatics.

Predicting protein function can be done in several different ways. The simplest is by direct similarity searching. If a new protein has a sequence that is nearly identical to that of a known protein, then their functions are likely to be similar too. If the entire new protein sequence is not identical to a known protein, there may be some portion of it that is similar to a conserved functional domain that has been found in many other proteins. Searching for a conserved sequence (a consensus with known variants) can be considerably more sensitive than direct similarity searching. The 3-D structure of the protein might be predicted and this structure then compared with other protein structures with known functions. Another approach relies on gene expression data. Genes that are expressed

Bioinformatics
By Stuart M. Brown

during a particular cellular activity might be ascribed a function related to that activity, or genes that share expression patterns with other better known genes may also share functions with those genes.

FINDING GENES IN DNA SEQUENCES

Gene finding is the identification of protein coding regions within DNA sequences, also known as open reading frames (ORFs). This is one of the single biggest challenges facing the bioinformatics specialists working in the Genome Project laboratories. Existing software is only 80% to 90% accurate in predicting genes in large stretches of genomic DNA—with both false positive and false negative errors. This may seem like an acceptable error rate for a small laboratory searching for a single gene on a 20 kb cosmid, but it is clearly not reasonable to allow 10% to 20% incorrect genes to go into GenBank® or to miss another 10% to 20%. This problem is exacerbated in eukaryotic genomes by the common occurrence of pseudogenes that are highly similar in sequence to real genes, but are not actually transcribed. As a result of these unsolved problems, there has been a proliferation of software designed to find genes in genomic DNA.

There are several ways to approach this problem. The simplest might be to do a similarity search against the expressed sequence tag (EST) databases or a translation and similarity search against the protein databanks (SWISS-PROT/TREMBL and GenPept are the most comprehensive). Both Basic Local Alignment Search Tool (BLAST) and FASTA offer automatic translate and search functions known as BLASTX and TFASTA, respectively (Figure 2). If a protein or EST sequence matches, align it with your unknown genomic sequence. The start and stop codons should line up nicely and the introns should also be obvious.

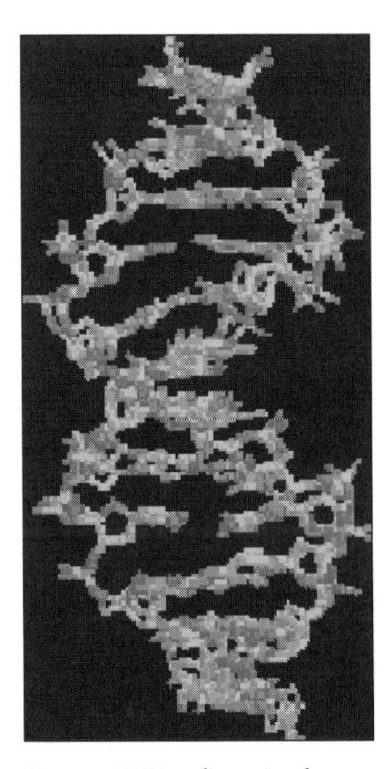

Figure 1. DNA 3-dimensional structural model. (See color plate A6.)

If you cannot find a handy template sequence in the databanks to identify genes in genomic DNA, then you must rely on knowledge of the DNA code. The transcription initiation site is generally an ATG codon, and it is usually about 30 bp downstream from a TAATAA sequence (or some close approximation). This is enough information to specify a pattern for a simple pattern search program. It may be even easier to produce just a graphic map of ORFs in all 6 reading frames and look for a long one. Simple software that maps an ORF starting at every ATG and stops it at every stop codon is available in a wide variety of forms. Genetics Computer Group (GCG) provides the FRAMES program, GeneWorks, MacVector™ (both from Oxford Molecular Group, Oxford, England, UK), and Sequencher (Gene Codes, Ann Arbor, MI, USA) all handle this function quite elegantly. Even venerable old DNA Strider (42) does a fine job of locating putative ORFs. Introns can be identified with moderate reliability by the occurrence of consensus splice signal sequences. A much better way to

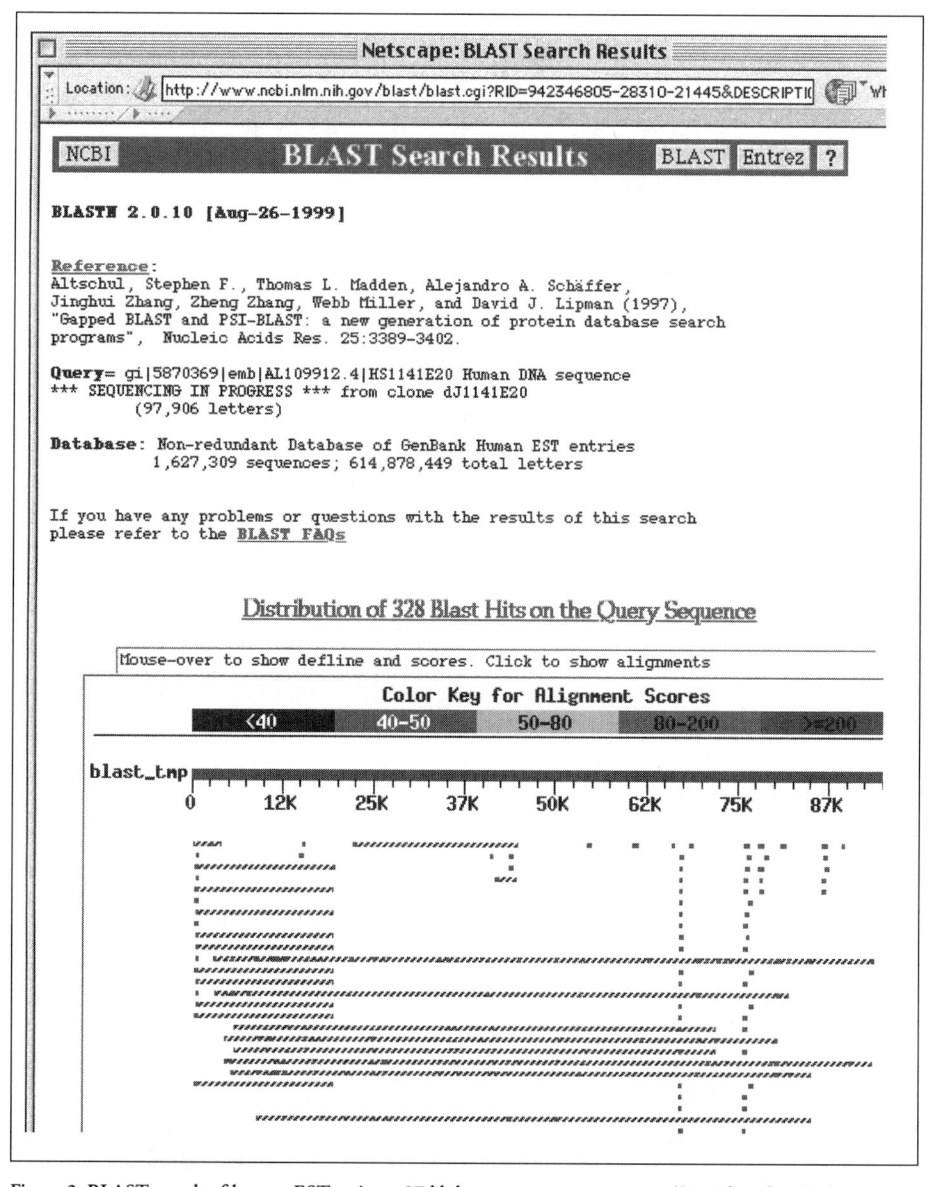

Figure 2. BLAST search of human ESTs using a 97 kb human genome sequence. (See color plate A7.)

spot introns is to align a cDNA sequence against a genomic sequence and look for large gaps. However, the only way to truly prove the existence of an intron is by experimentally comparing RNA (cDNA) to genomic sequences.

There are several other methods for identifying ORFs in DNA sequences. Several tools are available that help to identify protein coding sequences by statistics that measure codon usage and the non-random use of particular nucleotides in the third position of each codon. These statistical methods are imprecise, but can help identify possible coding regions in large chunks of genomic DNA sequence. There are other forms of custom software designed to identify ORFs, each using slightly different combinations of these pattern finding and statistical tools coupled, in some cases, with some form of similarity searching. The problem with gene finding programs is that they are not perfect. Every program will miss some ORFs, identify some false ORFs, and incorrectly identify the start and stop positions for exons. Whatever tool you use, the results will have to be checked by hand, and the only truly positive identifications of protein coding regions are those backed up by experimental evidence from the laboratory. With that in mind, here is a list of several popular tools for finding genes in genomic DNA.

GRAIL: Oak Ridge National Laboratory, Oak Ridge, TN, USA (58).
(**http://avalon.epm.ornl.gov/Grail-bin/EmptyGrailForm**)

ORFfinder: National Center for Biotechnology Information (NCBI).
(**http://www.ncbi.nlm.nih.gov/gorf/gorf.html**)

DNA translation: University of Minnesota Medical School.
(**http://alces.med.umn.edu/webtrans.html**)

GenLang (12).
(**http://www.cbil.upenn.edu/genlang/genlang_home.html**)

BCM GeneFinder: Baylor College of Medicine, Houston, TX, USA (53).
(**http://dot.imgen.bcm.tmc.edu:9331/seq-search/gene-search.html**)
(**http://dot.imgen.bcm.tmc.edu:9331/gene-finder/gf.html**)

DNA SIGNAL SEQUENCES

Once a stretch of DNA has been identified as a putative protein coding region —by similarity to a cDNA or protein, statistical resemblance to protein coding DNA, or just by the presence of start and stop codons—it is possible to gather confirming evidence by looking for sequences that regulate transcription. These regulatory sequences, which may be known as promoters, enhancers, transcription factors, or more generally as signal sequences, generally occur near open reading frames. A large number of these signal sequences have been identified and collected into databases. The best of these databases is called TransFac: The Transcription Factor Database (35) maintained by the German Gesellschaft für Biotechnologische Forschung mbH (GBF). TransFac currently has 4342 entries (as of October, 1999) with known protein binding and transcriptional regulatory functions. Another database of DNA signal sequences is the Eukaryotic Promoter Database (EPD), which is maintained by ISREC (Lausanne, Switzerland) as a subset of the European Molecular Biology Laboratory (EMBL) (48). The EPD provides a nearly comprehensive compilation of eukaryotic transcription signals (promoters). All information is directly abstracted from scientific literature. EPD has 1314 entries as of October, 1999.

Several tools make use of these databases to search DNA sequences for the occurrence of known signal sequences. The GCG program FINDPATTERNS can be used with the data file TFDATA (a local copy of the TransFac database). Also, the Signal Scan program (50), developed by Dan S. Prestridge (for both the Macintosh® and personal computer [PC]), can do the same thing on a copy of the TransFac (or EPD) database that you keep on your own machine—this has the advantage that you can very easily add your own new sequences to be searched.

There are many promoter finding programs available on the Web.

Promoter Scan; National Institutes of Health (NIH) Bioinformatics (BIMAS).
(**http://bimas.dcrt.nih.gov/molbio/proscan/**)

Signal Scan; NIH Bioinformatics (BIMAS).
(**http://bimas.dcrt.nih.gov:80/molbio/signal/index.html**)

Promoter Scan II; University of Minnesota and Axyx Pharmaceuticals.
(**http://biosci.cbs.umn.edu/software/proscan/promoterscan.htm**)

Search TransFac at GBF with MatInspector, PatSearch, and FunSiteP.
(**http://transfac.gbf-braunschweig.de/TRANSFAC/programs.html**)

The Computational Biology and Informatics Laboratory at the University of Pennsylvania offers a nice service called TESS (Transcription Element Search Software).
(**http://agave.humgen.upenn.edu/tess/index.html**)

TargetFinder; Telethon Institute of Genetics and Medicine, Milan, Italy (37).
(**http://hercules.tigem.it/TargetFinder.html**)

RNA STRUCTURE

Once DNA has been transcribed into RNA, an entirely different set of structural rules apply. RNA follows the same basic rules of base-pairing as DNA (with the substitution of uracil for thymine), but short single-stranded RNA molecules are much freer to take on a variety of 3-D shapes. This is particularly evident in the structures of tRNA molecules (Figure 3) and ribozymes as well as in the phenomenon of self-splicing of mRNA molecules.

Once again, the primary structure (the sequence) of the RNA molecule contains all of the information for self-assembly in complex 3-D structures. The RNA sequence must also contain: *(i)* the genetic code specifying the order of amino acids in proteins; *(ii)* the information that controls the beginning and ends of protein coding sequences (translation start and stop signals) and the splicing of introns; and *(iii)* information that determines the stability of the RNA molecule in the cell and its relative transcriptional level.

The process of self-assembly of tRNA molecules into their characteristic stem-loop structures is relatively well understood. A set of rules based on the concept of free energy minimization has been generated from the study of tRNAs that can be applied to any RNA sequence. From this algorithm Michael Zuker (59) created the computer program FoldRNA, which is available free for most types of computers. This program has also been incorporated into the GCG package and many other full-featured molecular biology programs. These tools allow many different views of RNA secondary structures, but do not necessarily predict the single optimal structure for a given sequence.

PROTEIN STRUCTURE

Proteins are polypeptides—long linear chains composed of mixed polymers of 20 amino acids. These linear polymers fold upon themselves to generate a shape characteristic of each different protein. In turn, this shape, along with the different chemical properties of the 20 amino acids, determines the function of the protein. This basic shape is highly conserved in evolution. Proteins with similar functions in distantly related organisms retain the same 3-D structure even if the individual amino acids have diverged so far that no significant sequence similarity can be detected.

Protein molecules are self-assembling, which means that all of the informa-

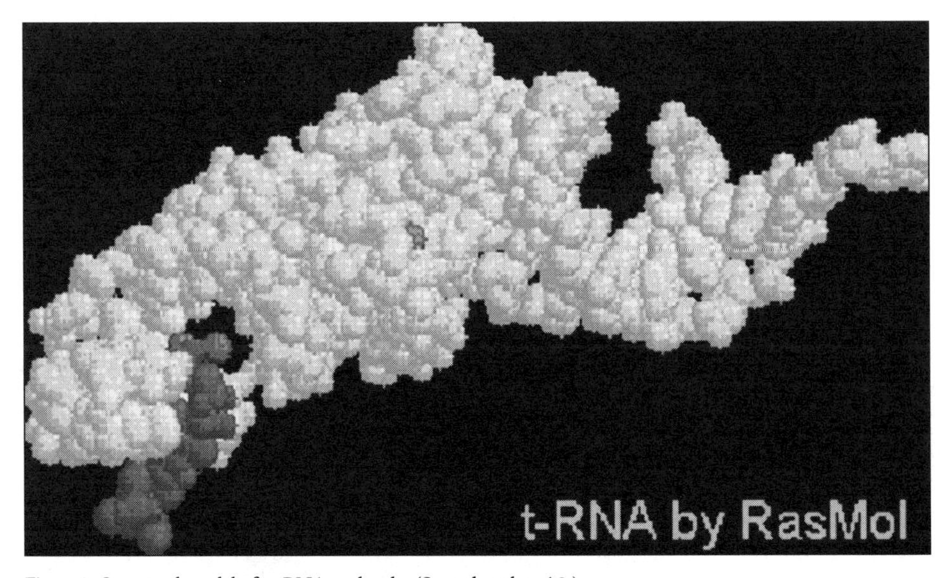

Figure 3. Structural model of a tRNA molecule. (See color plate A8.)

tion necessary to determine the final 3-D structure of the molecule is encoded in its sequence. This is known as Anfinsen's dogma (17). Thus, in theory, if one knows the sequence of a protein (the order with which the amino acids occur), one could infer its function.

Protein sequences are often described as having primary, secondary, and tertiary structure. The primary structure is the sequence itself, the order of amino acids. The secondary structure refers to local structural elements such as hairpins, helixes, β-pleated sheets, etc. However, most proteins fold into different secondary structures along different portions of their length. Tertiary structure is the final shape of the complete molecule such as a globular enzyme or a linear actin molecule. Proteins are occasionally referred to as having a quaternary structure that refers to complex structures composed of multiple subunits, each of which is a distinct polypeptide (encoded by its own gene) or even structures composed of mixtures of protein, DNA, and/or RNA (such as histones or RNA polymerase II).

The structure of a protein is produced by the folding of a peptide chain back on itself, and in some cases, the association of multiple peptide chains. This folding can occur as rotation around bonds within the constituent amino acids, as well as the bonds that join the amino acids one to another. The number of possible folding patterns for even a small polypeptide chain is effectively infinite.

The holy grail of bioinformatics is the ability to predict tertiary structure—and thereby function—from sequence information. It is not an exaggeration to state that the ability to exactly predict protein structures and, from that, protein function would revolutionize medicine, pharmacology, chemistry, and ecology. Clearly, we are not there yet, but there are some useful tools available. In many cases, local secondary structures can be predicted from the chemical properties of the amino acids. Certain amino acids in certain patterns tend to form helixes or β-sheets, and certain patterns of charged residues are also indicative of secondary structures such as coiled coils or membrane spanning regions.

The prediction of tertiary structure is currently more an art than a science. Without performing extensive X-ray crystallography studies, the only tools available to the computational biologist are comparison of new sequences with those of known structure and rule-based algorithms that attempt to minimize free energy. This is done using either Monte-Carlo methods or Neural Net software running on mainframes or supercomputers—not accessible tools for the average molecular biologist. The best tools for the prediction of protein structure are now a combination of homology and molecular modeling methods known as threading. First, a new protein is compared to the sequences of all proteins with known structures using standard similarity methods (52). Then the structures of the most similar proteins are used as templates for a molecular model of the new protein. Finally, the model is optimized using simulations of free energy and other known folding properties of individual amino acids.

The Protein Data Base (PDB) contains all of the known protein structures, which have been determined using x-ray crystallography and NMR. These structures are available as simple text file coordinates that can be used by appropriate software to create a 3-D molecular model of the protein. A wide variety of tools have been developed to view these structure files and manipulate the molecular images. The simplest of these software tools is CN3D, which was created by C.W. Hogue (33) and popularized by the NCBI. This was initially incorporated as part of the NENTREZ stand-alone client program used on PCs to browse the ENTREZ database, but now it is available as a plug-in for the Netscape and Internet Explorer Web browsers. The beauty of the CN3D program is that you can go to the ENTREZ Web site, locate some proteins of interest using the standard query tools, and then, if a protein has a known structure in the PDB, a link loads the structure file and automatically activates the CN3D viewer. The viewer not only lets you look at a 3-D image of the protein, but also allows rotation on any axis and change between spacefilling, ball and stick, and ribbon molecular models.

There are several other PDB viewers that also work as Web browser plug-ins:

1. RASMOL, a free program created by Roger Sayle (51), is very similar to CN3D, but it works with Netscape/Explorer as a helper application rather than as a plug-in (Figures 4 and 5).
2. CHIME, the CHIME plug-in, is a free Web browser plug-in created by MDL Information Systems, (8) that lets scientists display chemically significant 2-D and 3-D structures within a HyperText Markup Language (HTML) page or table. CHIME is scriptable and can be used to create interactive tutorials on any subject that can be enhanced by animated 3-D images.

   http://www.mdli.com/support/chime/
   http://www.umass.edu/microbio/chime/
3. Swiss-PDB Viewer was originally created by Nicolas Guex and Manuel C. Peitsch, Geneva Biomedical Research Institute. The Swiss-PDB Viewer is an application (for the Macintosh, PCs, and UNIX®) now distributed by the Swiss Institute of Bioinformatics (Geneva, Switzerland). It provides a user-friendly interface that allows the analysis of several proteins at the same time (28). The proteins can be superimposed in order to deduce structural alignments and compare their active sites or any other relevant parts. Amino acid mutations, H-bonds, angles, and distances between atoms are easy to obtain, thanks to the intuitive graphic and menu interface. It is possible to thread a protein primary sequence onto a 3-D template and get an immediate feedback of how well the threaded protein will be accepted by the reference structure. POV-Ray scenes can also be generated in order to make stunning ray-traced quality images.
4. There is a free program (for the Macintosh), called FoldIt, created by Jean-Claude Jesior, Centre National de la Recherche Scientifique (CNRS), Grenoble, France (34). FoldIt is a molecular modeling program to visualize and manipulate interactively protein structure files from the Brookhaven PDB including their hetero-atoms and water molecules. It is an integrated

environment in which statistical analysis as well as 3-D observations can be realized. Protein structures can be manipulated easily in real time with the mouse, zoomed, or observed in stereo. Animations can be created. Steric conflicts, disulfide bonds, hydrogen, and ionic interactions can be located

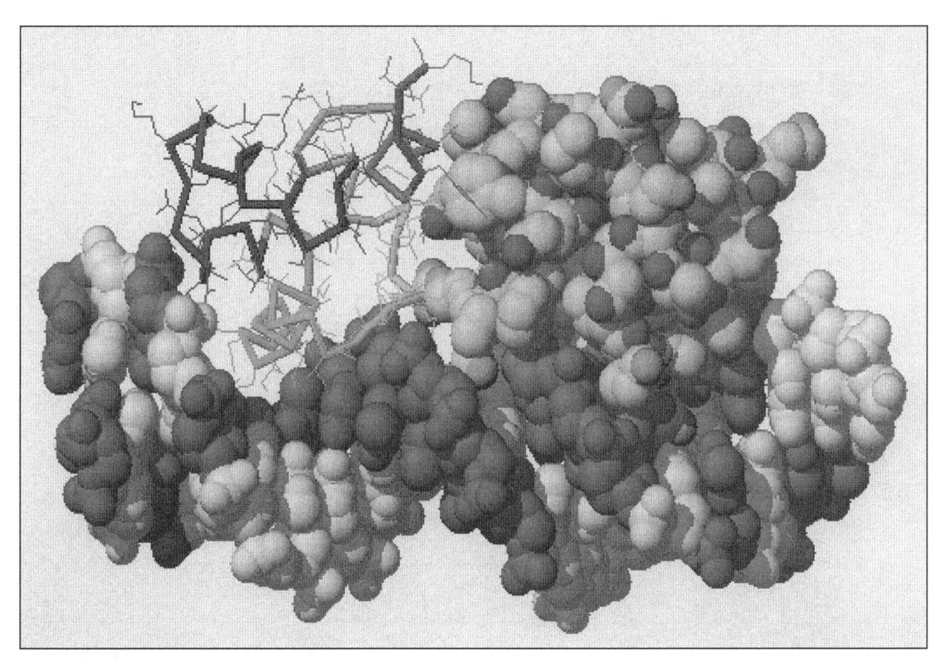

Figure 4. Phage CRO repressor bound to DNA; courtesy of Andrew Coulson and Roger Sayles with RasMol, University of Edinburgh, 1993. (See color plate A8.)

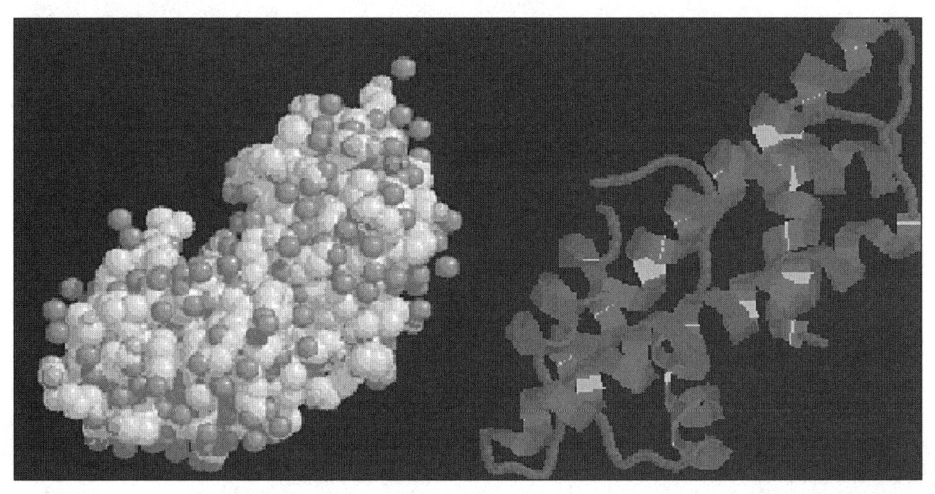

Figure 5. Histone molecule shown by RasMol in space filling and ribbon models. (See color plate A9.)

and displayed in the protein structure. Atoms and residues can be tagged individually (or globally), and structural information can then be extracted and reported in a text window. Pictures can be saved to the disk.

5. For PCs running Windows®, the ANTHEPROT package (24) permits the analysis of a protein sequence and 3-D structure analysis in a single interactive program. It also allows the user to:
   - View and handle 3-D protein structures from PDB files.
   - Study physico-chemical properties: hydrophobicity, antigenicity, flexibility, solvent accessibility, and amphiphilicity.
   - Create matrix dot plots.
   - Make multiple alignments.
   - Search for motifs using PROSITE database (3).
   - Search for homologous protein: program FASTA (W. Pearson [46]).
   - Identify levels between several sequences.
   - Predict secondary structure: Chou and Fasman (9) and Garnier (23).
   - Analyze circular dichroism spectra.
   - Predict transmembranous regions and structural domains.
   - Use such tools as editor, file, printing, format, colors, etc.

PROTEIN COMPOSITION AND SECONDARY STRUCTURE

If your new protein does not have a known 3-D structure, then what can be done? Some simple analyses of amino acid composition can give a hint as to the possible function of a protein because, in the broadest sense, the biochemical properties of a protein are determined by its amino acids. Molecular weight, pH, and isoelectric point can be calculated easily from the primary sequence. Hydrophobicity is a property of groups of amino acids, so it is best viewed in an average hydrophobicity plot (36), which might indicate regions that are extended (hydrophilic) or tightly wound up (hydrophobic) and might also identify membrane spanning regions (Figure 6).

Tools for these simple analyses of protein sequence are widely available as free software, as parts of comprehensive molecular biology packages, and on the Web at:

Protein Hydrophobicity Server; Bioinformatics Unit, Weizmann Institute of Science, Israel.

(http://bioinformatics.weizmann.ac.il/hydroph/)

SAPS (statistical analysis of protein sequences) (6): composition, charge, hydrophobic and transmembrane segments, cysteine spacings, repeats, and periodicity.

(http://www.isrec.isb-sib.ch/software/SAPS_form.html)

At the next level of complexity, protein molecules have local secondary structures that take the form of an α-helix, β-sheet, or a turn. Computational approaches to predicting these structures generally rely on the models of relationships between just a few amino acids, so predictions are done in a small window that moves down the protein (Figure 7). The essence of most of these tools is to assign each amino acid in the protein to one of three structural categories: α-helix, β-sheet, or hairpin turn (also called a β-loop or β-turn). This assignment is based on data about the chemical properties of each amino acid and its immediate neighbors (9). Unfortunately, these predictions are only 60% to 70% accurate for average proteins (21), and different methods will give different results (23). The results are a prediction for each amino acid of whether it is most likely helix, sheet or turn. However, none of these algorithms are very accurate, and the use of several different algorithms will often give different answers. Better secondary structure predictions can be made from a multiple alignment of a group of related protein sequences (20).

There are a wide variety of tools available on the Web that perform protein secondary structure predictions. Careful judgment is required in the interpretation of output from custom algorithms and experimental Web servers.

SSPRED is a three state secondary structure prediction routine utilizing the algorithm described in Mehta et al. (43). Polypeptides are compared to the entire SWISS-PROT database, and based on the structure of similar proteins, predictions are made for local regions that are assigned to one of three structures: helical, strand, or coil/loop regions.

(http://www.embl-heidelberg.de/sspred/sspred_info.html)

STRIDE recognizes secondary structural elements in proteins from their atomic coordinates (19). It utilizes both hydrogen bond energy and main chain dihedral angles. It relies on database-derived recognition parameters with the crystallographers' secondary structure definitions as a standard-of-truth.

(http://www.embl-heidelberg.de/argos/stride/stride_info.html)

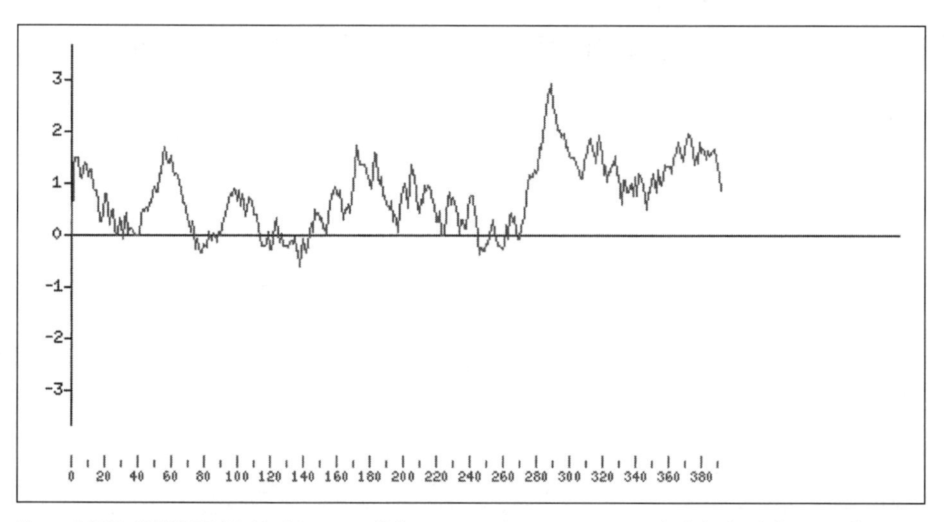

Figure 6. P53_HUMAN (P04637) human cellular tumor antigen p53, Kyte-Doolittle hydrophilicty, window = 19.

Secondary Structural Content Prediction (SSCP); EMBL, Heidelberg, Germany (16).
(http://www.bork.embl-heidelberg.de/SSCP/sscp_seq.html)
BCM Search Launcher; Protein Secondary Structure Prediction; Baylor College of Medicine (53).
(http://dot.imgen.bcm.tmc.edu:9331/seq-search/struc-predict.html)
PREDATOR; EMBL, Heidelberg, Germany.
The mean prediction accuracy of PREDATOR is 68% for a single sequence and 75% for a set of related sequences (20,21). The gain in prediction accu-

```
                   ....,....1....,....2....,....3....,....4....,....5....,....6
        AA        |MMSGAPSATQPATAETQHIADQVRSQLEEKYNKKFPVFKAVSFKSQVVAGTNYFIKVHVG|
        PHD sec   |             HHHHHHHHHHHHHHHH       EEEEEEEEEEEEE EEEEEEE   |
        Rel sec   |999997899667599999999998999765587784336888999999923339999658|
detail:
        prH sec   |000000000221289999999998999876201111100000000000000000000000|
        prE sec   |000000000000000000000000000010000023578889989888536699999720|
        prL sec   |999898889777600000000001000112688886531111000000036330000278|
subset: SUB sec   |LLLLLLLLLLLLHHHHHHHHHHHHHHHLLLLL...EEEEEEEEEE....EEEEEELL|


ACCESSIBILITY
3st:    P_3 acc   |bbebbeeeeeebbeebbebbeebeeebeeeeeee eebebbebebbbbbb bbbbeb bb|
10st:   PHD acc   |007006778670077007007706760777777737707007060000000500006050 0|
        Rel acc   |103021343252044604644672424555547615444425212186671016926120|
subset: SUB acc   |.......e..e..eeb.ebbeeb.e.beeeeeee.eebeb.e....bbbb...bb.b...|
```

Figure 7. Secondary structure prediction for a short amino acid sequence.

racy compared to other techniques is achieved by better accounting for long-range effects, utilization of homologous information in the from of carefully selected pairwise alignment fragments, and reliance on a much larger collection of protein primary structures.

(**http://www.embl-heidelberg.de/cgi/predator_serv.pl**)

University of California at Los Angeles, Department of Energy (UCLA-DOE) Protein Fold Recognition Server.

(**http://www.doe-mbi.ucla.edu/people/fischer/TEST/getsequence.html**)

Super-Secondary Structure

In addition to the standard helix, sheet, and turn structures that apply to all proteins, it is possible to recognize some structural motifs such as membrane spanning domains (transmembrane domains) (16,17), coiled coils (41), helix-turn-helix (12), and signal peptides (45) that are present in some proteins. These simple structural motifs are sometimes known as super-secondary structures. They are not present in all proteins, but if they are present, they give important hints about protein function. Programs for identifying these structural motifs are included in GCG and some other molecular biology packages, and at a wide variety of Web servers.

The PredictProtein server provides multiple sequence alignments and predictions of secondary structure, residue solvent accessibility, and the location of transmembrane helices.

(**http://www.embl-heidelberg.de/predictprotein/**)

SOSUI; discrimination of membrane proteins from soluble ones, prediction of transmembrane helical regions (31), Tokyo University of Agriculture and Technology, Japan.

(**http://azusa.proteome.bio.tuat.ac.jp/sosui/**)

TMpred (transmembrane prediction); Kay Hofmann (32), ISREC (Swiss Institute for Experimental Cancer Research).

(**http://www.ch.embnet.org/software/TMPRED_form.html**)

Dense Alignment Surface (DAS) Transmembrane Prediction Server. DAS is based on low-stringency dot plots of the query sequence against a collection of nonhomologous membrane proteins using a previously derived, special scoring matrix (11).

(**http://www.biokemi.su.se/~server/DAS/**)

COILS (coiled coil prediction); ISREC (41).

(**http://www.ch.embnet.org/software/COILS_form.html**)

Helix-turn-helix motif prediction (12), Network Protein Sequence @nalysis at Pôle Bio-Informatique Lyonnais, Universite Claude Bernard, Lyon, France

(**http://pbil.ibcp.fr/cgi-bin/npsa_automat.pl?page=/NPSA/npsa_hth.html**)

SignalP (signal peptides); signal peptide prediction, Technical University of Denmark (45).
(**http://www.cbs.dtu.dk/services/SignalP/**)
The PredictProtein server provides multiple sequence alignments, detection of functional motifs, and predictions of secondary structure, residue solvent accessibility, as well as the location of transmembrane helices and coiled coils.
(**http://www.embl-heidelberg.de/predictprotein/**)

PROTEIN TERTIARY STRUCTURE

At the present time, a protein's full 3-D structure cannot be accurately predicted from sequence alone (this is known as *ab initio* structure prediction). Levinthal's paradox states that the number of possible 3-D conformations of a protein increases as an exponential function of the number of amino acids present (38). So a 100 amino acid protein has 3^{200} possible backbone configurations, which is many orders of magnitude beyond the capacity of the fastest computers—in fact this computation would take a theoretical 100 GFLOP (billion floating point operations per second) supercomputer longer than the current age of the universe. So it seems to be theoretically impossible for a computer to ever be able to make a brute force computation of protein structure by examining every possible structure. On the other hand, many scientists believe that there are perhaps only a few hundred basic protein substructures, but we do not yet have this structural vocabulary defined, nor the ability to recognize variants on a theme.

Instead of brute force structure prediction from sequence alone, protein structure bioinformatics has emphasized comparisons of new sequences with other sequences of known structure using a process called threading. Threading uses the basic backbone of a known structure, attempts to match a new sequence to that backbone, and then makes adjustments based on known thermodynamic principles that affect polypeptides in solution. Unfortunately, threading only works well if the protein of unknown structure has at least approximately 25% sequence similarity with a protein of known structure. Also, the current state-of-the-art in threading requires complex custom software, an expert user, and several days of computing time on a high-powered workstation. Nevertheless, some Web sites do offer quick approximate threading predictions. Threading algorithms are receiving intensive effort from may researchers, and these approaches will become more powerful as the database of known protein structures grows.

UCLA-DOE Protein Fold Recognition Server (threading) (18).
(**http://www.doe-mbi.ucla.edu/people/fischer/TEST/getsequence.html**)
SwissModel; ExPASy, University of Geneva, Switzerland (28,47).
(**http://www.expasy.ch/swissmod/SWISS-MODEL.html**)
CPHmodels; Technical University of Denmark (40).
(**http://www.cbs.dtu.dk/services/CPHmodels/**)

MOTIFS

Proteins are built out of functional units known as domains or motifs. These domains have sequences that are conserved over time and across species, often much more similar than other regions in a protein. The exon splicing theory of Walter Gilbert (25) postulated that exons correspond to protein folding domains, which in turn serve as functional units. In some cases, proteins that have entirely different functions may share a single similar exon such as an ATPase region or a DNA binding domain.

Known protein motifs have been collected into several databases. The best starting point for a protein motif search is the PROSITE database: a Dictionary of Protein Sites and Patterns maintained by Amos Bairoch at the University of Geneva, Switzerland (3). PROSITE contains a comprehensive list of documented protein domains constructed by expert molecular biologists. Each domain in PROSITE is defined by a simple pattern and a list of proteins known to have that domain (a protein family) (Figure 8). Patterns can have a defined set of alternate amino acids at each position, and defined spaces, but no gaps. For example this is the PROSITE entry for an ATP/GTP binding motif: [AG]-x(4)-G-K-[ST].

Comparisons of a query sequence against PROSITE patterns is done by exact matching, so new variants of known motifs may not be found. It is possible to make a pattern search with mismatches, but false positives can quickly accumulate if too many mismatches are allowed.

Several simple tools have been developed to compare an amino acid sequence to the patterns in PROSITE. GCG provides the program MOTIFS. Several different stand-alone programs are available for Macintosh, Windows PCs, and most other forms of computers. The PROSITE database is sufficiently small (a few megabytes) so that a diligent user could download it by file transfer protocol (FTP) onto a PC and make searches of protein sequences with these programs that take only a minute or so to complete. It is also possible to create your own patterns and use these programs to search local protein databases for matches.

MacPattern is an excellent free program developed by Rainer Fuchs (22) for pattern searching using PROSITE and other databases on the Macintosh.
(**ftp://ftp.ebi.ac.uk/pub/software/mac/macpattern.hqx**)

PATMAT is a free DOS program by Wallace and Henikoff (56) that will work on most Windows PCs.
(**ftp://ncbi.nlm.nih.gov/repository/blocks/patmat.dos**)

A number of Web servers are also available for on-line scanning of your query sequence for the occurrence of PROSITE patterns.

ScanProsite at ExPASy; University of Geneva, Switzerland.
(**http://www.expasy.ch/sprot/scnpsit1.html**)

Network Protein Sequence Analysis (NPSA); Institut de Biologie et Chimie des Protéines, Lyon, France.
(**http://pbil.ibcp.fr/NPSA/npsa_prosite.html**)

PPSRCH; European Bioinformatics Institute (EBI), Cambridge, England, UK.
(http://www2.ebi.ac.uk/ppsearch/)

One cautionary note: some of the amino acid patterns in PROSITE are quite short and occur very frequently in many different proteins. Some of these occurrences are simply random chance, while others represent motifs, such as glycosylation sites, that really do occur quite frequently in proteins. A search of your protein sequence with the complete PROSITE database is quite likely to reveal some matches to these common motifs, which may be completely without statistical or functional significance.

PROFILES

Pattern searching has some serious drawbacks for detecting motifs in proteins. Since the patterns are designed by hand using a specified set of known proteins, new variants of domains are not likely to be found by pattern matching. It would be much more useful to search for sequences that are *similar* to known domains rather than look for exact matches to predefined patterns. Profile searching is a method of applying the rules of similarity to protein domains.

Profiles are tables of amino acid frequencies at each position in a motif (26). A profile is built from a multiple alignment of conserved domains from the members of a protein family. This table can then be used for similarity searches (using algorithms like BLAST and FASTA) of individual proteins or against a database. If additional members of a protein family are found, they can be added to the

```
PROSITETOGCG of: Prosite.Doc and Prosite.Dat December 18, 1995 11:18
Release 13.0  (11/95)

Name Offset Pattern ..  PDoc_Name

11s_Seed_Storage 1 NGx(D,E)2x(L,I,V,M,F)C(S,T)x{11,12}(P,A,G)D  0284.pdoc
1433_1 1 RNL(L,I)SV(G,A)YKN(I,V)      0633.pdoc
1433_2 1 YK(D,E)STLIMQLL(R,H)DNLTLW(T,A)(S,A)  0633.pdoc
25a_Synth_1  1 GGSx(A,G)(K,R)xTxL(K,R)(G,S,T)xSD(A,G)    0653.pdoc
25a_Synth_2  1 RPVILDPx(D,E)PT    0653.pdoc

/////////////////////////////////////////////////////////////////

Zinc_Finger_C2h2 1 Cx{2,4}Cx3(L,I,V,M,F,Y,W,C)x8Hx{3,5}H 0028.pdoc
Zinc_Finger_C3hc4 1 CxHx(L,I,V,M,F,Y)Cx2C(L,I,V,M,Y,A) 0449.pdoc
Zinc_Protease 1 (G,S,T,A,L,I,V,N)x2HE(L,I,V,M,F,Y,W)~(D,E,H,R,K,P...
Zn2_Cy6_Fungal 1 (G,A,S,T,P,V)Cx2C(R,K,H,S,T,A,C,W)x2(R,K,H)x2Cx{5..
Zp_Domain 1 (L,I,V,M,F,Y,W)x7(S,T,A,P,D,N)x3(L,I,V,M,F,Y,W)x( ...
```

Figure 8. PROSITE patterns.

multiple alignment for that motif, and the profile can be re-computed to improve the sensitivity of future searches. Profile searching is more sensitive than pattern searching and it allows gaps.

The PROSITE database (3) includes profiles that have been computed from each protein family. This concept has been extended by a number of other databases that have attempted to categorize all known proteins into families, or at least cluster them by shared domains using various algorithms based on the concept of profiles. These automatically computed databases contain more domains than PROSITE, but they may lack annotation about the function of the domains. All of these protein motif/domain/family databases are available for searching on the Web.

PROSITE ProfileScan; ExPASy, Geneva.
(http://www.isrec.isb-sib.ch/software/PFSCAN_form.html)

BLOCKS; The Henikoff group at the Fred Hutchinson Cancer Research Center, Seattle, WA, USA. Builds profiles from PROSITE entries and adds all matching sequences in SWISS-PROT (29,30).
(http://www.blocks.fhcrc.org/blocks_search.html)

PRINTS; Terri Attwood at the UCL in London (1,2). Builds profiles from automatic alignments of OWL nonredundant protein databases. PRINTS contains 614 entries, encoding 3280 individual motifs. Of these entries, 302 have some sort of equivalent pattern in PROSITE.
(http://www.biochem.ucl.ac.uk/cgi-bin/fingerPRINTScan/fps/PathForm.cgi)

ProDom from Daniel Kahn at the Institut National de la Recherche Agronomique in Toulouse, France, has a nice graphical interface that allows the user to see the domain structure of proteins and download multiple alignments (Figure 9) (10,54).
(http://www.toulouse.inra.fr/prodom/doc/blast_form.html)

Pfam is a collection of protein family alignments, some of which are constructed semiautomatically using hidden Markov models (HMMs) and some are fully automatically constructed (1344 protein family HMM profiles built by hand) (5,55).
(http://pfam.wustl.edu/hmmsearch.shtml)

SBASE; The International Centre for Genetic Engineering and Biotechnology (ICGEB). Entries are clustered using the BLAST score as a similarity measure (contains 7231 entries) (44,49).
(http://www2.icgeb.trieste.it/~sbasesrv/)

SCOP; Structural Classification of Proteins (based on the PDB). As of October, 1998 (Release 1.41) SCOP contains 7467 PDB entries classified into 16 211 domains (4,39).
(http://scop.stanford.edu/scop/)

Any study of a new protein or a family of known proteins would do well to begin by thoroughly investigating the contents of these protein structure/domain databases. Much work (database searching, multiple alignment, and the building of phylogenetic trees) has already been done—be sure that your efforts build on this knowledge rather than repeating these same analyses.

SEARCHING FOR NEW MOTIFS

In some cases, rather than search your sequence for the known motifs found in a database, you might wish to search a database with a motif of your own creation. Motif searching may discover homologies between short portions of genes (perhaps functional domains) that cannot be found using traditional similarity search programs like BLAST or FASTA.

It is possible to use any of the pattern searching tools designed to work with PROSITE patterns with a text string of DNA or amino acids of your own design. Each program has different requirements for formatting the pattern to allow for ambiguities such as enclosing different choices in parentheses, including "X" characters that match anything, etc.

The GCG program FINDPATTERNS is a powerful and flexible pattern searching tool. FINDPATTERNS works best with short patterns (less than about 50 characters). A number of patterns can be searched simultaneously by creating a pattern list file. All of the other Mac, PC, and Web-based programs listed above for PROSITE pattern searching can also be used with custom patterns.

All of the reasons that make profiles more powerful than patterns for searching for new members of known protein families also apply in searches with new motifs. There are several tools for creating custom profiles for new protein motifs. These tools all require a multiple alignment of sequences containing the motif. There is no specific limitation on the size of the multiple alignment region or on the number of sequences in the alignment. However, an alignment confined to a highly conserved region with only similar sequences will make a better profile. Searches with profiles that contain too much ambiguity will match too many sequences in a database.

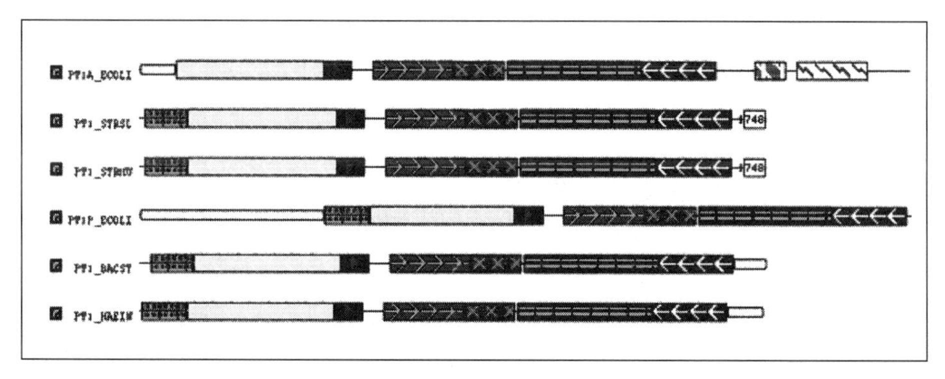

Figure 9. Example of ProDom domains. (See color plate A9.)

GCG contains a very good set of profile analysis programs based on the original work on profiles by Gribskov et al. (26). Philipp Bucher of ISREC (Swiss Institute for Experimental Cancer Research) has created a simple stand-alone version of these profile tools called PFTOOLS (7). Sean Eddy at Washington University, St. Louis, MO, has created the HMMER program (15), which achieves a similar result using HMMs (14). Unfortunately, these programs are quite demanding of computer power, and they are not currently available for free on the Web. HMMER and PFTOOLS are both freely available as source code to load on your own SGI (Silicon Graphics, Mountain View, CA, USA), Sun Sparc (Sun Microsystems, Palo Alto, CA, USA), or Compaq Alpha workstation (Compaq Computer Corp., Houston, TX, USA).

In some cases, you may suspect that a group of protein (or DNA) sequences share some common motif, but you are unable to identify it in a multiple alignment. For example, several seemingly unrelated proteins may be bound by another protein in a yeast two-hybrid experiment, or a major histocompatibility complex (MHC) molecule may bind several antigens. Alternately, several genes may be co-regulated due to some stimulus, so a common promoter or enhancer might be present in 5′ or 3′ noncoding DNA, or a regulatory sequence may be present in the mRNA, or a sequence in the protein itself may be a target for posttranslational regulation. All of these situations are good starting material for the discovery of patterns in unaligned sequences. The MEME program at the San Diego Supercomputing Center provides a HMM-based tool to discover these patterns (Figure 10) (27):

(**http://www.sdsc.edu/MEME/meme/website/meme.html**)

One downside of MEME is that it always finds some common patterns in any group of sequences—regardless of whether that pattern has any true biological significance. Unfortunately, MEME does not provide a good set of statistical tools for investigators to evaluate the quality of the motifs that it discovers. MEME does produce an e-value similar to BLAST when it compares its motif against the set of starting sequences, but since the pattern was found in the original set of sequences, they will always match the motif with a highly significant

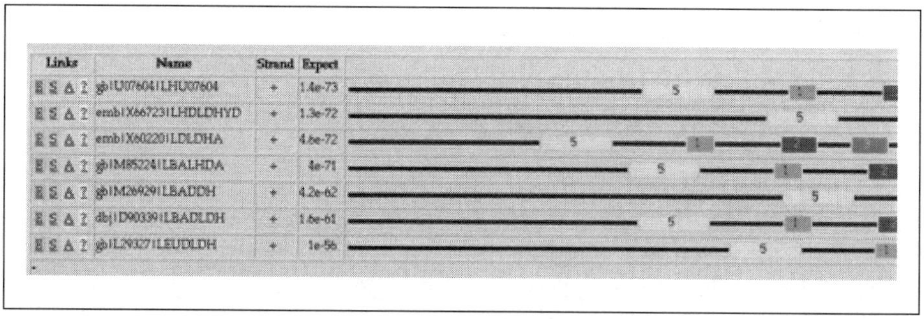

Figure 10. A sample MEME output. (See color plate A9.)

e-value. MEME does produce nice-looking colored graphs of the motifs located on the original sequences.

REFERENCES

1. **Attwood, T.K. and M.E. Beck.** 1994. PRINTS—a protein motif fingerprint database. Protein Eng. *7*: 841-848.
2. **Attwood, T.K., M.D. Croning, D.R. Flower, A.P. Lewis, J.E. Mabey, P. Scordis, J.N. Selley, and W. Wright.** 2000. PRINTS-S: the database formerly known as PRINTS. Nucleic Acids Res. *28*:225-227
3. **Bairoch, A.** 1993. The PROSITE dictionary of sites and patterns in proteins, its current status. Nucleic Acids Res. *21*:3097-3103.
4. **Barton, G.J.** 1994. Scop: structural classification of proteins. Trends Biochem. Sci. *19*:554-555.
5. **Bateman, A., E. Birney, R. Durbin, S.R. Eddy, K.L. Howe, and E.L. Sonnhammer.** 2000. The Pfam Protein Families Database. Nucleic Acids Res. *28*:263-266
6. **Brendel, V., P. Bucher, I.R. Nourbakhsh, B.E. Blaisdell, and S. Karlin.** 1992. Methods and algorithms for statistical analysis of protein sequences. Proc. Natl. Acad. Sci. USA *89*:2002-2006.
7. **Bucher, P., K. Karplus, N. Moeri, and K. Hofmann.** 1996. A flexible search technique based on generalized profiles. Comput. Chem. *20*:3-23.
8. **CHIME**, distributed as free non-commercial software by the author: MDL Information Systems, San Leandro, CA (**http://www.mdli.com/support/chime/**).
9. **Chou, P.Y. and G.D. Fasman.** 1978. Prediction of the secondary structure of proteins from their amino acid sequence. Adv. Enzymol. *47*:45-148.
10. **Corpet, F., F. Servant, J. Gouzy, and D. Kahn.** 2000. ProDom and ProDom-CG: tools for protein domain analysis and whole genome comparisons. Nucleic Acids Res. *28*:267-269.
11. **Cserzo, M., E. Wallin, I. Simon, G. von Heijne, and A. Elofsson.** 1997. Prediction of transmembrane alpha-helices in procariotic membrane proteins: the dense alignment surface method. Protein Eng. *10*:673-676.
12. **Dodd, I.B. and J.B. Egan.** 1990. Improved detection of helix-turn-helix DNA-binding motifs in protein sequences. Nucleic Acids Res. *18*:5019-5026.
13. **Dong, S. and D.B. Searls.** 1994. Gene structure prediction by linguistic methods. Genomics *23*:540-551.
14. **Durbin, R., S. Eddy, A. Krogh, and G. Mitchison.** 1998. Biological Sequence Analysis: Probabilistic Models of Proteins and Nucleic Acids. Cambridge University Press, Cambridge, UK.
15. **Eddy, S.** 1998. HMMER. Biological sequence analysis using profile hidden Markov models. Version 2.1.1. Source code distributed freely by the author under the terms of the GNU General Public License (**http://hmmer.wustl.edu/**).
16. **Eisenhaber, F., F. Imperiale, P. Argos, and C. Frommel.** 1996. Prediction of secondary structural content of proteins from their amino acid composition alone. I. New analytic vector decomposition methods. Proteins *25*:157-168.
17. **Epstein, C.J., R.F. Goldberger, and C.B. Anfinsen.** 1963. The genetic control of tertiary protein structure: studies with model systems. Cold Spring Harb. Symp. Quant. Biol. *28*:439-449.
18. **Fischer, D. and D. Eisenberg.** 1996. Protein fold recognition using sequence-derived predictions. Protein Sci. *5*:947-955.
19. **Frishman, D. and P. Argos.** 1995. Knowledge-based protein secondary structure assignment. Proteins *23*:566-579.
20. **Frishman, D. and P. Argos.** 1996. Incorporation of non-local interactions in protein secondary structure prediction from the amino acid sequence. Protein Eng. *9*:133-142.
21. **Frishman, D. and P. Argos.** 1997. Seventy-five percent accuracy in protein secondary structure prediction. Proteins *27*:329-335.
22. **Fuchs, R.** 1994. Predicting protein function: a versatile tool for the Apple Macintosh. Comput. Appl. Biosci. Apr; *10*:171-178.
23. **Garnier, J., D.J. Osguthorpe, and B. Robson.** 1978. Analysis and accuracy of simple methods for predicting the secondary structure of globular proteins. J. Mol. Biol. *120*:97-120.
24. **Geourjon, C. and G. Deleage.** 1995. ANTHEPROT 2.0: a three-dimensional module fully coupled with protein sequence analysis methods. J. Mol. Graph. *13*:209-212.
25. **Gilbert, W.** 1978. Why genes in pieces? Nature *271*:501.
26. **Gribskov, M., R. Luthy, and D. Eisenberg.** 1990. Profile analysis. Methods Enzymol. *183*:146-159.
27. **Grundy, W.N., T.L. Bailey, C.P. Elkan, and M.E. Baker.** 1997. Meta-MEME: motif-based hidden Markov models of protein families. Comput. Appl. Biosci. *13*:397-406.
28. **Guex, N. and M.C. Peitsch.** 1997. SWISS-MODEL and the Swiss-PDB Viewer: an environment for comparative protein modeling. Electrophoresis *18*:2714-2723.
29. **Henikoff, S. and J.G. Henikoff.** 1991. Automated assembly of protein blocks for database searching. Nucleic Acids Res. *19*:6565-6572.

30. **Henikoff, J.G., E.A. Greene, S. Pietrokovski, and S. Henikoff.** 2000. Increased coverage of protein families with the blocks database servers. Nucleic Acids Res. *28*:228-230.
31. **Hirokawa, T., S. Seah Boon-Chieng, and S. Mitaku.** 1998. SOSUI: classification and secondary structure prediction system for membrane proteins. Bioinformatics *14*:378-379.
32. **Hofmann, K. and W. Stoffel.** 1993. TMBASE—a database of membrane spanning protein segments. Biol. Chem. Hoppe Seyler *374*:166.
33. **Hogue, C.W.** 1997. CN3D: a new generation of three-dimensional molecular structure viewer. Trends Biochem. Sci. *22*:314-316.
34. **Jésior, J.C., A. Filhol, and D. Tranqui.** 1994. FOLDIT(LIGHT)—an interactive program for Macintosh computers to analyze and display protein data bank coordinate files. J. Appl. Cryst. *27*:1075.
35. **Knuppel, R., P. Dietze, W. Lehnberg, K. Frech, and E. Wingender.** 1994. TRANSFAC retrieval program: a network model database of eukaryotic transcription regulating sequences and proteins. J. Comput. Biol. *1*:191-198.
36. **Kyte, J. and R.F. Doolittle.** 1982. A simple method for displaying the hydropathic character of a protein. J. Mol. Biol. *157*:105-132.
37. **Lavorgna, G., A. Guffanti, G. Borsani, A. Ballabio, and E. Boncinelli.** 1999. TargetFinder: searching annotated sequence databases for target genes of transcription factors. Bioinformatics *15*:172-173.
38. **Levinthal, C.** 1969. How to fold graciously. *In* J.T.P. DeBrunner and E. Munck (Eds.), Mossbauer Spectroscopy in Biological Systems (Proceedings of a Meeting Held at Allerton House, Monticello, Illinois). University of Illinois Press, Urbana, IL.
39. **Lo Conte, L., B. Ailey, T.J. Hubbard, S.E. Brenner, A.G. Murzin, and C. Chothia.** 2000. SCOP: a structural classification of proteins database. Nucleic Acids Res. *28*:257-259.
40. **Lund, O., K. Frimand, J. Gorodkin, H. Bohr, J. Bohr, J. Hansen, and S. Brunak.** 1997. Protein distance constraints predicted by neural networks and probability density functions. Protein Eng. *10*:1241-1248.
41. **Lupas, A., M. Van Dyke, and J. Stock.** 1988. Predicting coiled coils from protein sequences. Science *252*:1162-1164.
42. **Marck, C.** 1988. DNA Strider: a C program for the fast analysis of DNA and protein sequences on the Apple Macintosh family of computers. Nucleic Acids Res. *16*:1829-1836.
43. **Mehta, P.K., J. Heringa, and P. Argos.** 1995. A simple and fast approach to prediction of protein secondary structure from multiply aligned sequences with accuracy above 70%. Protein Sci. *4*:2517-2525.
44. **Murvai, J., K. Vlahovicek, E. Barta, B. Cataletto, and S. Pongor.** 2000. The SBASE protein domain library, release 7.0: a collection of annotated protein sequence segments. Nucleic Acids Res. *28*:260-262.
45. **Nielsen, H., J. Engelbrecht, S. Brunak, and G. von Heijne.** 1997. Identification of prokaryotic and eukaryotic signal peptides and prediction of their cleavage sites. Protein Eng. *10*:1-6.
46. **Pearson, W.R. and D.J. Lipman.** 1998. Improved tools for biological sequence comparison. Proc. Natl. Acad. Sci. *85*:2444-2448.
47. **Peitsch, M.C.** 1996. ProMod and Swiss-Model: Internet-based tools for automated comparative protein modelling. Biochem. Soc. Trans. *24*:274-279.
48. **Perier, R.C., T. Junier, C. Bonnard, and P. Bucher.** 1999. The eukaryotic promoter database (EPD): recent developments. Nucleic Acids Res. *27*:307-309.
49. **Pongor, S., V. Skerl, M. Cserzo, Z. Hatsagi, G. Simon, and V. Bevilacqua.** 1993. The SBASE domain library: a collection of annotated protein segments. Protein Eng. *6*:391-395.
50. **Prestridge, D.S.** 1991. SIGNAL SCAN: a computer program that scans DNA sequences for eukaryotic transcriptional elements. Comput. Appl. Biosci. *7*:203-206.
51. **Sayle, R.A. and E.J. Milner-White.** 1995. RASMOL: biomolecular graphics for all. Trends Biochem. Sci. *20*:374.
52. **Smith, T.F., and M.S. Waterman.** 1981. Identification of common molecular subsequences. J. Mol. Biol. *147*:195-197.
53. **Smith, R.F., B.A. Wiese, M.K. Wojzynski, D.B. Davison, and K.C. Worley.** 1996. BCM search launcher—an integrated interface to molecular biology data base search and analysis services available on the World Wide Web. Genome Res. *6*:454-462.
54. **Sonnhammer, E.L. and D. Kahn.** 1994. Modular arrangement of proteins as inferred from analysis of homology. Protein Sci. *3*:482-492.
55. **Sonnhammer, E.L., S.R. Eddy, and R. Durbin.** 1997. Pfam: a comprehensive database of protein domain families based on seed alignments. Proteins *28*:405-420.
56. **Wallace, J.C. and S. Henikoff.** 1992. PATMAT: a searching and extraction program for sequence, pattern and block queries and databases. Comput. Appl. Biosci. *8*:249-254.
57. **Watson, J.D. and F.H. Crick.** 1974. Molecular structure of nucleic acids: a structure for deoxyribose nucleic acid. Nature *248*:765.
58. **Xu, Y., R. Mural, M. Shah, and E. Uberbacher.** 1994. Recognizing exons in genomic sequence using GRAIL II. Genet. Eng. (NY) *16*:241-253.
59. **Zuker, M.** 1989. Computer prediction of RNA structure. Methods Enzymol. *180*:262-288.

8 Plasmid Mapping and PCR Primer Design

RESTRICTION MAPPING

Making restriction maps is a routine laboratory activity that is necessary for any type of cloning project. In addition, maps are a common way for laboratories to archive information about entire libraries of plasmid constructs. For archival purposes, it is important that map data are stored in a reliable format that is broadly supported in the bioinformatics community, so that the obsolescence of a particular computer program does not render archives unusable. High-quality maps are important for publications and exchange of information between researchers or between laboratories.

Restriction mapping is a computationally simple problem. There is an extremely wide variety of software available for both personal computers (PCs) and mainframes ranging from simple text-based freeware to expensive and sophisticated graphics applications with database management and project history functionality. If all you need to know is "Where are the *Bam*HI sites in pUC18?", then virtually any software will suffice. If you need to print out publication-quality circular maps of your vector construction strategy, then a custom application for a PC would probably be your best bet. I am not aware of any first-rate mapping applications available from free Web servers.

GCG Mapping Tools

MAP is the Genetics Computer Group (GCG) restriction enzyme mapping program. Like a lot of GCG programs, it is very powerful and quite complex. By default, MAP includes protein translations (in three forward reading frames). This can be changed to any, all six, or none. Restriction sites can be mapped for all enzymes (the default) or any enzymes that you specify by name. The output can be viewed as a linear map or in a table format (with the -TABLE option). You can also select enzymes that have 6 base or larger recognition sites (six-cutters in

Bioinformatics
By Stuart M. Brown

laboratory slang) with the parameter -MINS = 6, just blunt end cutters (-OVERHANG = 0) or just 5′ and/or 3′ overhangs (-OVERHANG = 5, -OVERHANG = 3, or -OVERHANG = 5,3) (Figure 1).

MAP offers a large number of other customization options that include:

1. Treating a sequence as circular (-CIR).
2. Allowing for mismatches between an enzyme's recognition site and the corresponding site in your test sequence (-MIS).
3. Limiting the output to enzymes that cut just once (-ONCE), at least two (or any number of) times (-MINCUTS = 2), no more than two (or any number of) times (-MAXCUTS = 2), or those that do not cut at all (-NONCUT).
4. Specific regions of your test sequence can be excluded from the enzyme search.
5. The -SILENT option allows you to search for places where a single base mutation can create a recognition site for a specific enzyme.

MAPPLOT uses the same algorithm (and the same options) as MAP, but creates a graphical output designed for a plotter. These output files can be saved in GCG's FIGURE format, transferred to a Macintosh® by file transfer protocol (FTP), and then viewed and printed with GCG's free Mac program, GCGFigure. It can also create a text form of its output file suitable for printing on an ordinary laser printer.

MAPSORT simulates a restriction digest and allows you to predict the sizes of the digest products with any combination of enzymes. It uses essentially the same set of options as MAP and MAPPLOT.

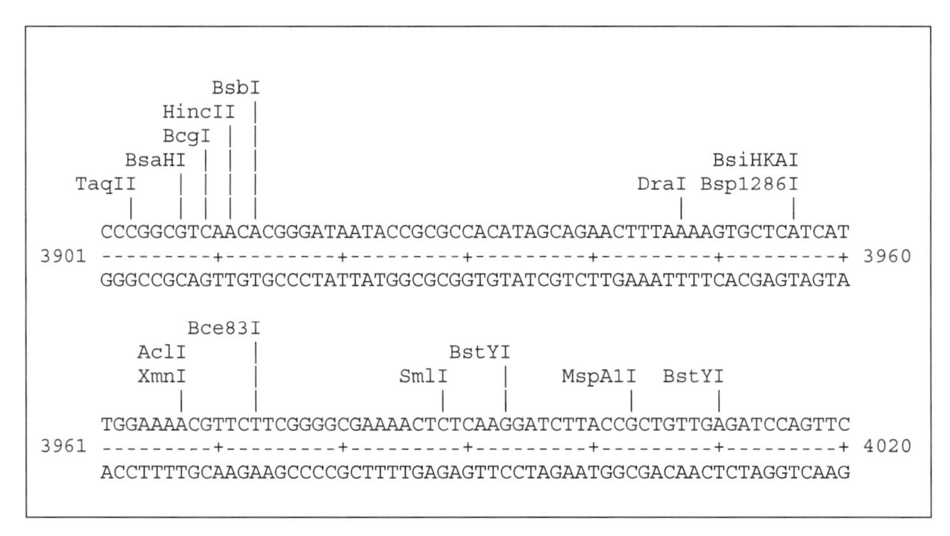

Figure 1. Part of a map of pBR322 created with GCG MAP.

FINDPATTERNS allows you to search a sequence for short patterns such as restriction enzyme recognition sites, promoter binding sites, etc. If you are looking for a restriction site that is not in the list available to MAP (i.e., not in REBASE) or for a variant on a restriction site that cannot be specified with the -MISMATCH or -SILENT options of MAP, then try FINDPATTERNS.

PLASMIDMAP is a GCG program that produces a publication-quality circular map of a plasmid construct (Figure 2). It uses input files generated by MAPSORT and a text file that contains data about blocks and ranges within a sequence to create output in the GCG FIGURE format. The use of PLASMIDMAP is too complex to explain here. This program has great power and can be used to produce very elegant and intricate figures, but it is not for the GCG novice (or anyone with little patience for the peculiarities of GCG).

DNA STRIDER

One of my favorite old DNA analysis programs, DNA Strider (6), produces very neat, simple restriction maps with a minimum of fuss—both as annotated sequence and as graphic maps (Figure 3). The maps can be saved in Macintosh PICT format, which allows for editing of each text and graphical element in a drawing program such as Canvas® (Deneba Software, Miami, FL, USA),

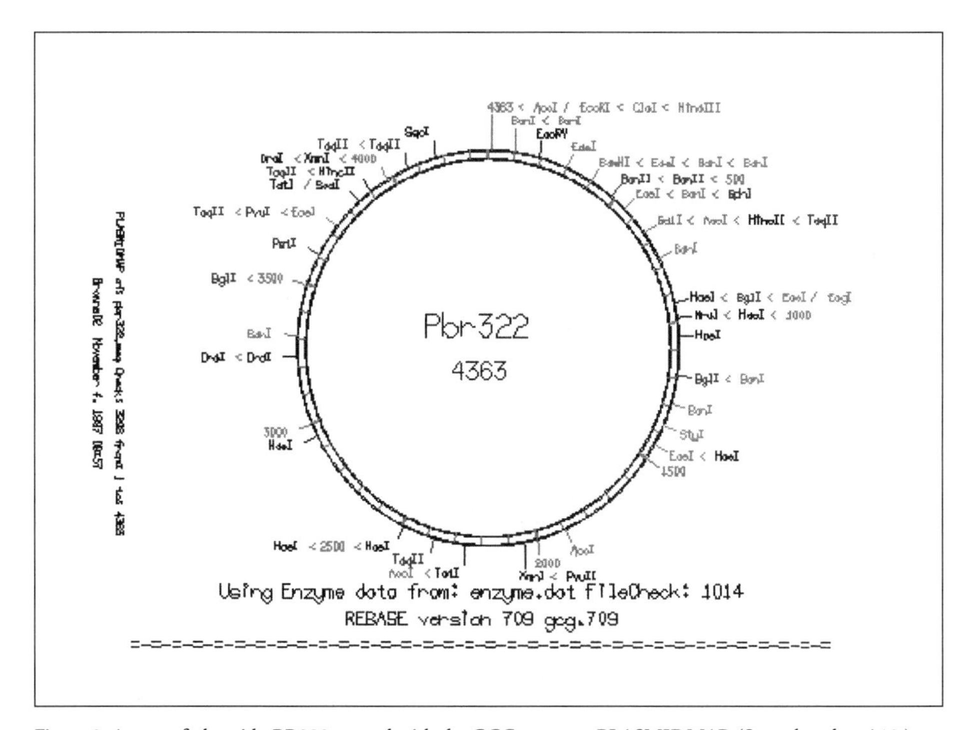

Figure 2. A map of plasmid pBR322 created with the GCG program PLASMIDMAP. (See color plate A10.)

Illustrator® (Adobe Systems, Seattle, WA, USA), or Freehand® (Macromedia, San Francisco, CA, USA). The program also does a good job of translating DNA to protein (1, 3, or 6 reading frames) and basic protein analysis such as hydrophobicity, acid/base map, amphipathic prediction of α-helix and β-sheets.

DNA Strider versions 1.0 and 1.1 are in wide circulation among molecular biologists as freeware (although not currently available from any of the well-known software FTP archives). The current version of DNA Strider is version 1.3, which is available directly from the author, Christian Marck, for the bargain price of $200.

Christian Marck
Service de Biochimie - Bat 142
Centre d'Etudes Nucleaires de Saclay
91191 Gif-Sur-Yvette Cedex France

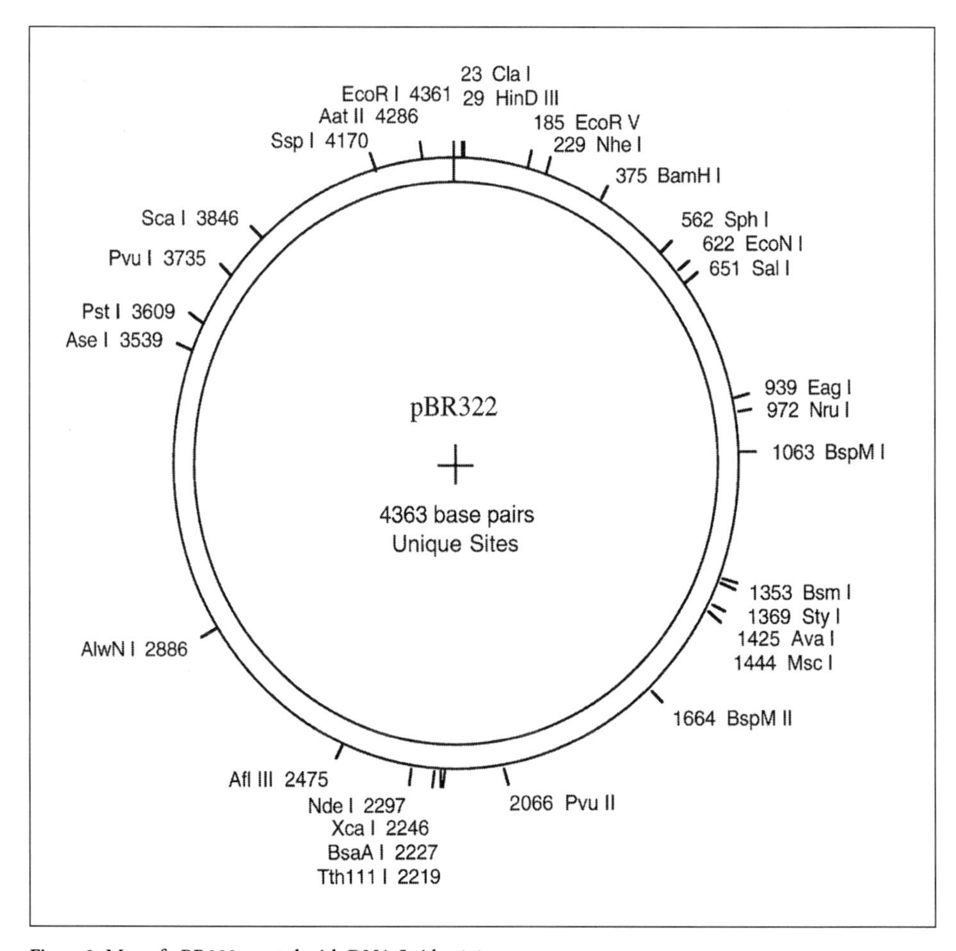

Figure 3. Map of pBR322 created with DNA Strider 1.1.

MACVECTOR AND OMIGA

MacVector® and OMIGA™ are comprehensive molecular biology programs for Macintosh and Windows PC computers, respectively, both currently owned and marketed by Oxford Molecular Group (Oxford, England, UK). Making and printing publication-quality restriction maps in MacVector and OMIGA is very simple (Figure 4). However, I do not recommend using these programs to archive data for plasmid constructs. The file formats used are proprietary, and there is no guarantee that these files will be readable a few years from now given the ever accelerating pace of change in the PC industry. In fact, the once popular GeneWorks® program, after moving among three different owners over 5 years, was officially discontinued by Oxford Molecular effective December 31, 1997. In contrast, GCG uses universal text and graphics file formats for all of its program output, so you can be reasonably sure that this data will remain readable for many years to come.

Features of MacVector include multiple alignment with CLUSTALW, protein analysis including motif searching and graphical displays of hydrophobicity, antigenicity, secondary structure prediction, and transmembrane helices. DNA analysis tools include restriction mapping, PCR primer design, linear or circular graphical

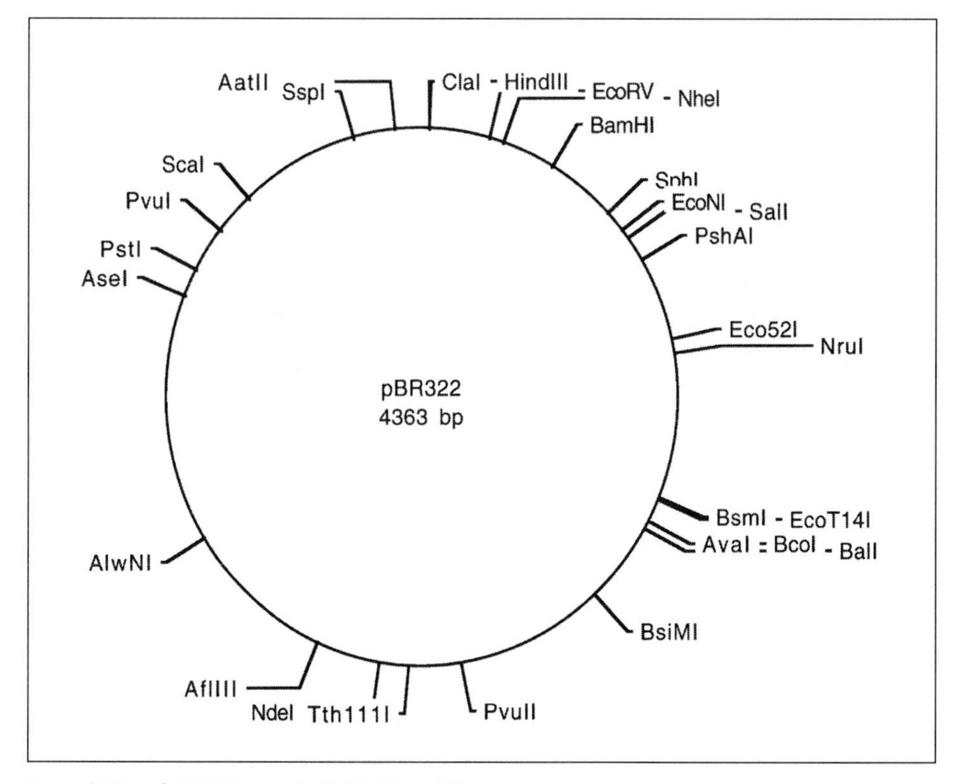

Figure 4. Map of pBR322 created with MacVector 5.0.

maps displaying cut sites and features, and open reading frame (ORF) detection.

Oxford Molecular Group has recently developed a new comprehensive molecular biology program called OMIGA for Windows 95® and Windows NT® computers. OMIGA has a similar set of features as Oxford's MacVector product for Macintosh computers (Figure 5). OMIGA version 2.0 (available in 2000) will also provide an interface to GCG running on a mainframe computer, so users can run large-scale database searches from within the program without needing to maintain large databases on their desktop computers.

GENE CONSTRUCTION KIT

The Gene Construction Kit (Textco, West Lebanon, NH, USA) is an extremely elegant and powerful plasmid drawing program (Figure 6). It is

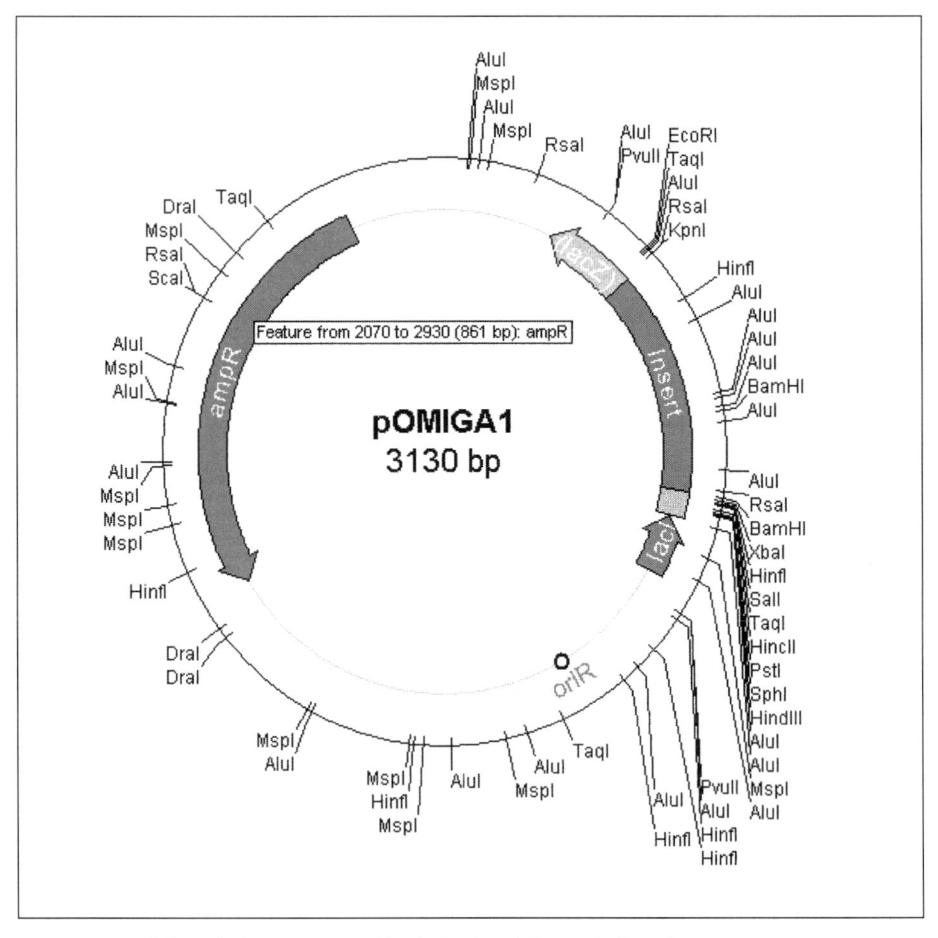

Figure 5. Map of plasmid pOMIGA1 created by OMIGA with features indicated.

designed to create plasmid constructs using simulated cloning procedures: input your sequence, find restriction sites, digest with an enzyme, add linkers, take a commercial vector cut with a restriction enzyme, ligate, cut the resulting construct with another enzyme, ligate in a promoter fragment derived from a digest of another plasmid, etc. It tracks all of your manipulations; for example, if a new vector is created by fusing several sequences in separate cloning experiments, each step can be viewed and arranged in a history window. There is also a Layout window, which allows you to arrange the pictures of your construct with text and additional graphics to produce publication-quality figures.

At $999 this is not a tool that every laboratory must have, but for those who work with plasmid construction every day, it will pay for itself quickly, both in time saved preparing laboratory notes and publications, as well as time and reagents saved from improved experimental designs.

VECTOR NTI

Vector NTI™ is a new program from InforMax (Bethesda, MD, USA) that has ambitions far beyond restriction mapping and drawing plasmid graphics

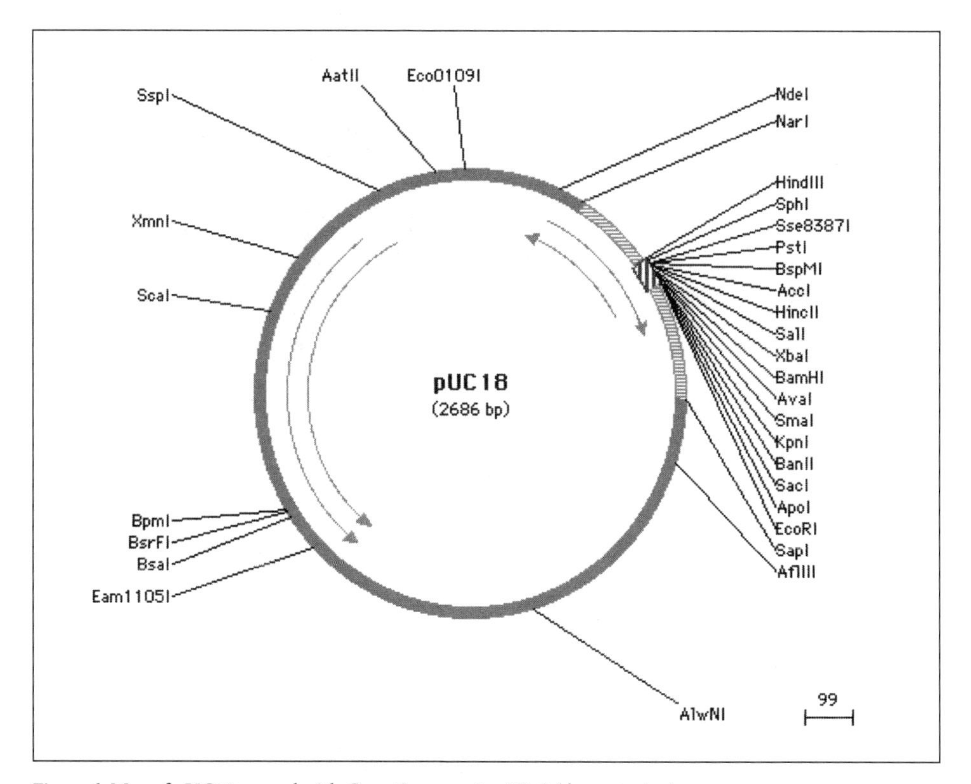

Figure 6. Map of pUC18 created with Gene Construction Kit 2 (demo version).

(Figure 7). This program is designed to be a central database of all of a laboratory's plasmid constructs, polymerase chain reaction (PCR) primers, and other oligonucleotides. It also includes PCR primer design functions as well as a rudimentary tool for accessing Web-based sequence analysis sites and importing the results of analyses from Web pages. Vector NTI can automatically read GenBank® and other sequence formats and create graphical maps of sequences that include all of the features. It is then easy to extract coding sequences from genomic DNA, simulate subcloning of promoters, etc.

The program takes an intelligent approach to plasmid construction. You can simulate the construction of a new molecule by adding inserts, combining pieces from several different plasmids or vectors, etc. You can also set preferences for which laboratory techniques you do or do not like (for example, you can tell the

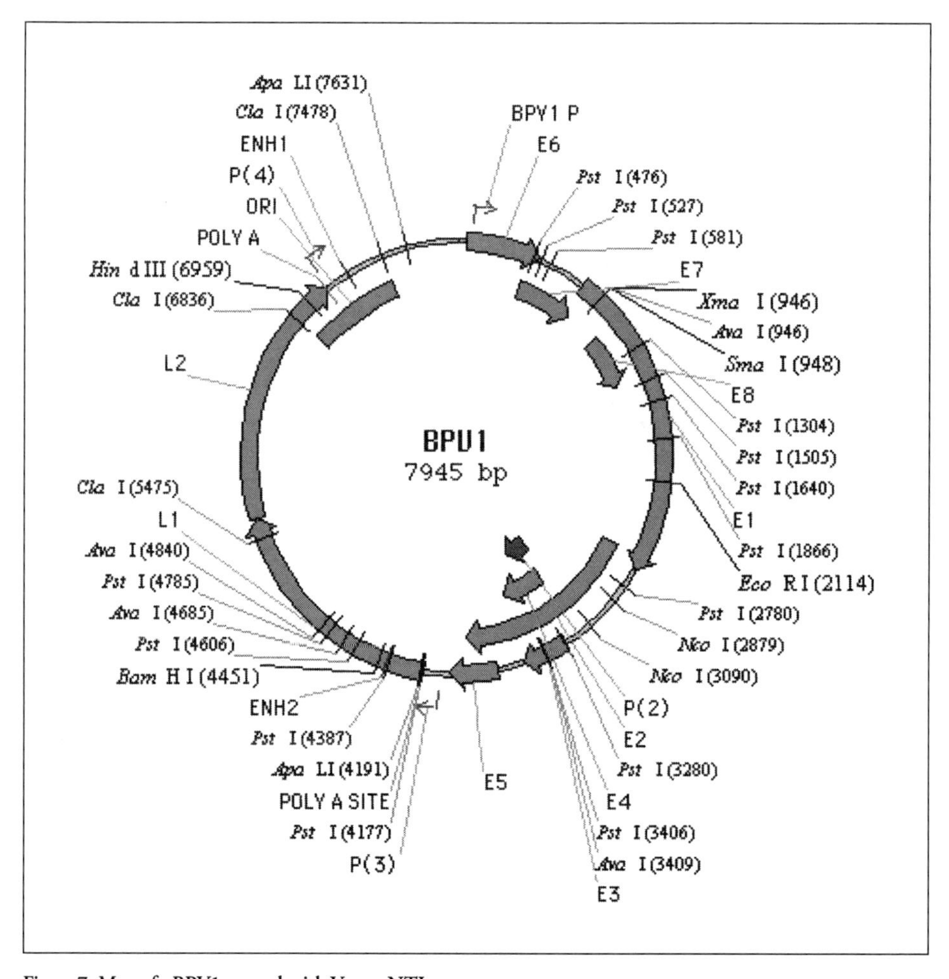

Figure 7. Map of pBPV1 created with Vector NTI.

program that you do not want to use blunt–blunt ligations). The software will even point out if a recombinant plasmid has lost a restriction site or gained a new site that the parent molecule did not have. The plasmid drawing tools are excellent, allowing for easy production of publication-quality graphics.

PRIMER DESIGN FOR PCR AND SEQUENCING

The design of PCR (and sequencing) primers is relatively simple from a computational point of view: just search along a sequence and find short subsequences that fit certain criteria. However, since the molecular biology of PCR is very complex, the nature of these criteria is not at all obvious. Rules for choosing PCR primers are a rough combination of educated guesses and old-fashioned trial-and-error.

Some general guidelines are: *(i)* primers should be at least 15 bp long, *(ii)* have at least 50% GC content, and *(iii)* anneal at a temperature in the range of 50° to 65°C. Usually higher annealing temperatures (T_m) are better (i.e., more specific for your desired target). In addition, the forward and reverse primer should anneal at approximately the same temperature (allowing perhaps 3° or 4°C of difference between them). However, none of the published formulas for calculating T_m has been proven to give better than a rough ballpark estimate (i.e., plus or minus about 5° or even 10°C), so how well can you trust an estimate of the difference between the T_m of two primers?

Specificity is of course related to the uniqueness of the region that you are attempting to amplify. Repeated sequences can be amplified, but only if unique flanking regions can be found where primers can bind. Primers that match multiple sequences will give multiple products.

Next, you have to consider the formation of self-annealing regions within each primer (i.e., hairpin and foldback loops) as well as direct annealing between two primers to form the dreaded primer dimers. It is also possible that primers will bind to regions within the sequence fragment that is being amplified.

Once you get into multiplex PCRs, with several sets of different primers working simultaneously in a single amplification tube making different products, no software can realistically predict what is going to happen. It is probably best to design each primer set on its own merits, keep all of the annealing temperatures within a narrow range, and just use trial-and-error to assemble multiplex sets.

PRIMER DESIGN PROGRAMS

There are a number of different software packages that can be used to create PCR (and sequencing) primers. They all provide essentially the same set of features, and any of them can be used to design primers that will work. Different research applications may benefit from particular algorithms or special features in one particular program.

GCG has a well-respected tool known simply as PRIME. A similar program known as PRIMER3 is available on a Web server at the Whitehead Institute, Massachusetts Institute of Technology (MIT) (7) (**http://www-genome.wi.mit.edu/genome_software/other/primer3.html**).

The GCG program PRIME is a good tool for the design of primers for PCR and sequencing. Like all GCG programs, it takes some effort to understand PRIME well enough to use it effectively. Selection of the basic parameters is quite simple and accomplished through a series of interactive questions (these parameters can also be set in the command line).

For PCR primer pair selection, you can choose a target range of the template sequence to be amplified. For DNA sequencing primers, you can specify positions on the template that must be included in the sequencing.

In selecting appropriate primers, PRIME allows you to specify a variety of constraints on the primer and amplified product sequences.

1. Upper and lower limits for primer and product melting.
2. Temperatures.
3. Primer and product GC contents.
4. A range of acceptable primer sizes.
5. A range of acceptable product sizes.
6. Required bases at the 3′ end of the primer (3′ clamp).
7. Maximum difference in melting temperatures between a pair of PCR primers.

PRIME uses a simulated annealing test to check individual primers for self-complementarity and to check the two primers in a PCR primer pair for complementarity to each other. Using this same annealing test, PRIME has the option to screen against nonspecific primer binding on the template sequence and on any repeated sequences that you specify.

There are a whole host of other primer design applications for every conceivable type of computer. The comprehensive molecular biology packages for PCs (MacVector, LaserGene™ [DNASTAR, Madison, WI, USA], OMIGA, etc.) all include simple and elegant PCR primer design functions. Many people find these programs to be easier to use than GCG for primer design.

There are a wide variety of other programs available. Free programs include Amplify (2), Primer (5), and PrimerDesign (DOS only). Commercial programs include Oligo® (Mac and PC; Molecular Biology Insights, Cascade, CO, USA), Designer PCR (PC), RightPrimer (Mac; BioDisk Software, San Francisco, CA, USA), and Primer Premier (Windows; Premier Biosoft International, Palo Alto, CA, USA).

There are also some Web sites that offer free PCR primer design functions.

Primers! hosted by Williamstone Enterprises.

(**http://www.williamstone.com/primers/index.html**)

Web Primer at the Saccharomyces Genome Database in the Department of

Genetics at the School of Medicine, Stanford University.
(**http://genome-www2.stanford.edu/cgi-bin/SGD/web-primer**)
GeneFisher (3) at Bielefeld University, Germany.
(**http://bibiserv.techfak.uni-bielefeld.de/genefisher/**)
Primer3 (7) at the Whitehead Institute/MIT Center for Genome Research, Cambridge, MA, USA.
(**http://www-genome.wi.mit.edu/cgi-bin/primer/primer3.cgi**)
xprimer (4) at the Virtual Genome Center, University of Minnesota Medical School.
(**http://alces.med.umn.edu/rawprimer.html**)

Primo at the University of Texas Southwestern Medical School is optimized for making sequencing primers for primer walking.
(**http://pompous.swmed.edu/compbio/primo**)

Most of these programs use variations on a similar algorithm and therefore will generally produce a similar set of primer recommendations for a given sequence. Among all of these programs, one stands out with some unusual features. RightPrimer screens primer sequences for uniqueness against a mathematical profile of GenBank subdivisions (the appropriate subdivision is chosen by the user). Rarer primers would be expected to be more selective for the target sequence against this species background. RightPrimer can also screen multiplex sets of primers for inter-primer interactions.

PCR primers generated by any of these programs can only be evaluated experimentally—by making the primers and performing some reactions under a range of PCR conditions. Performance of primers should be evaluated for several criteria:

1. specificity: synthesis of the desired product versus other nonspecific products.
2. optimal annealing temperature and Mg^{++} concentration: primers should function optimally within a temperature range and at Mg^{++} concentrations that work for your laboratory (i.e., similar to your other primers).
3. amount of product: primers that produce larger quantities of product with fewer numbers of PCR cycles are better.

There is little experimental evidence to support one program over another for producing better PCR primers. Every investigator will have to evaluate these tools for themselves. If you generate some data comparing primers designed with different software, publish it! Other scientists would love to see your results.

It is likely that the differences between primers selected by the various programs are due more to differences in default settings for various parameters or differences in user interface that lead to different user choices for parameters, rather than fundamental differences in algorithms. It is also possible that the differences between programs—whether due to differences in algorithms or in parameter settings—lead to primers that may be more effective for one type of experiment than another. Differences in the actual performance of primers may also be due to PCR conditions favored by a particular investigator or other peculiarities of a particular type of experiment. There probably is not a superior PCR primer design program that makes better primers for all types of reactions. Choices of software are more likely to be influenced by convenience, esthetics of a user interface, or an unscientific feeling that better primers are produced by one program versus another.

REFERENCES

1. **Becker, A., J. Napiwotzki, and M.S. Damian.** 1995. Primer Design—a new program to choose PCR primers and oligonucleotide probes. Medizinische Genetik *7*:A-158. Shareware, available from the authors: (**http://www.chemie.uni-marburg.de/~becker/pdhome.html**).
2. **Engels, B.** 1993. Contributing software to the Internet: the Amplify program. Trends Biochem. Sci. *11*:448-450. Freely available at (**http://www.wisc.edu/genetics/CATG/amplify/index.html**).
3. **Giegerich, R., F. Meyer, and C. Schleiermacher.** 1996. GeneFisher—software support for the detection of postulated genes. Ismb *4*:68-77.
4. **Griffais, R., P.M. Andre, and M. Thibon.** 1991. K-tuple frequency in the human genome and polymerase chain reaction. Nucleic Acids Res. 19:3887-3891.
5. **Lincoln, S., M. Daly, and E.S. Lander.** 1991. PRIMER: a computer program for automatically selecting PCR primers. The Whitehead Institute. Freely availalable from the authors (**http://www-genome.wi.mit.edu/ftp/distribution/primer.0.5**).
6. **Marck, C.** 1988. DNA Strider: a C program for the fast analysis of DNA and protein sequences on the Apple Macintosh family of computers. Nucleic Acids Res. *16*:1829-1836.
7. **Rozen, S. and H.J. Skaletsky.** 2000. Primer3 on the WWW for general users and for biologist programmers. Methods Mol. Biol. *132*:365-386. Code freely available at (**http://www-genome.wi.mit.edu/genome_software/other/primer3.html**).

9 DNA Sequencing

THE BIOCHEMISTRY OF DNA SEQUENCING

DNA sequencing technology is the source for much of the data used in bioinformatics. It is important for anyone doing sequence analysis to have a thorough understanding of how this data is collected in the laboratory. This is especially important for mathematicians and computer scientists whose daily tasks generally keep them out of touch with the laboratory. In fact, several months of experience working in a sequencing laboratory should be a prerequisite for anyone wishing to develop sequence analysis algorithms or software. There is no substitute for firsthand experience with the types of errors that can occur in the sequencing process and with the computer software needs of laboratory workers. The computer is an essential part of the sequencing process, but sequencing software must take into account every aspect of a sequencing project: project planning, cloning, the biochemistry of sequencing, methods for visualization of experimental results, data entry, sequence assembly, and analysis.

Virtually all modern DNA sequencing, regardless of whether it is done on an automated sequencer or manually, relies on the Sanger method of controlled interruption of DNA replication (5). This method uses a DNA polymerase enzyme to make many copies of a piece of DNA (the template). The polymerase enzyme requires a primer (a short single-stranded piece of DNA that exactly matches the template) as well as all four types of deoxynucleotides (G, A, T, and C) in order to synthesize new DNA strands. The sequencing reaction also includes some form of label (radioactive, fluorescent, or immunological) attached to the primer or to the nucleotides, so that the DNA products can be visualized.

The sequencing reaction mixture also contains a small amount of dideoxynucleotides which stop (terminate) the replication process if they are incorporated in the growing DNA chain instead of the normal deoxynucleotide bases, since they lack the 3′-OH needed for chain extension. For each template, four separate sequencing reactions are set up that contain one each of the dideoxynucleotides (G, A, T, and C) corresponding to the four DNA bases. For example, in the reaction mixture containing dideoxyadenosine triphosphate, some of the growing

Bioinformatics
By Stuart M. Brown

strands will terminate when they reach each T in the template sequence. The resulting set of DNA fragments form a nested set, all starting with the primer sequence, but ending at different T residues. Similar reactions are set up that stop replication at G, T, and C residues by including trace amounts of the respective dideoxynucleotide triphosphate. These fragments are separated by length using polyacrylamide gel electrophoresis; one gel lane for each of the four different dideoxynucleotide reactions (Figure 1). Then, the sequence is read off of the gel from the positions of the bands.

Real autoradiograms are somewhat more difficult to read than the diagram in Figure 1 (Figure 2). Problems such as background, uneven spacing of bands, uneven intensity of bands, compression of bands for particular sequences, uneven migration of lanes, curvature (smiling) of the gel, etc. all contribute to the difficulty of correctly reading the DNA sequence from an autoradiogram.

There are some computer tools that can aid in the interpretation of autoradiograms. A piece of equipment called a digitizer can be used to aid in the process of entering the bases from the autoradiogram into the computer. A digitizer uses

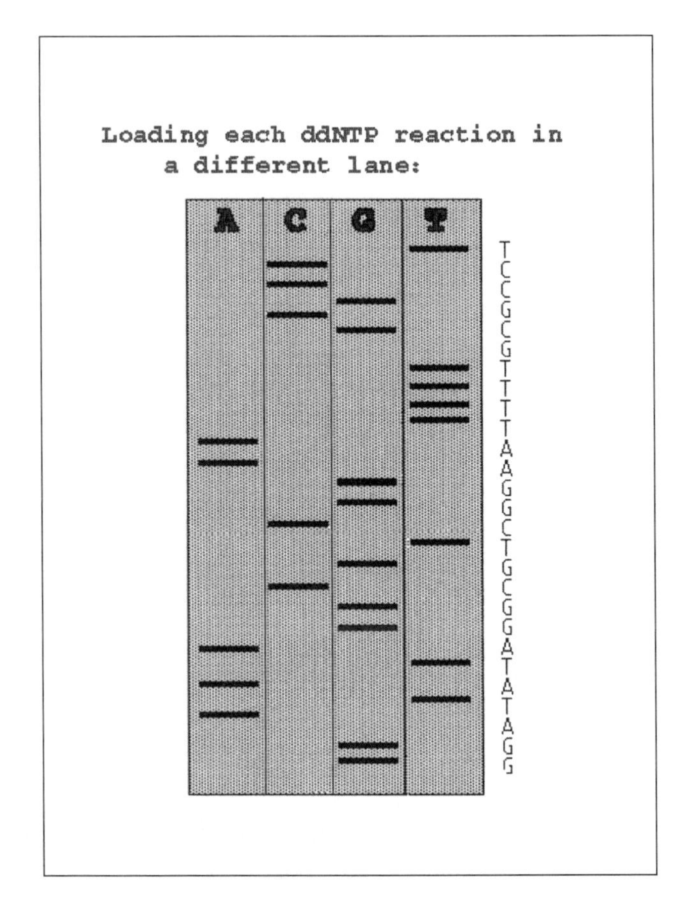

Figure 1. Diagram of a sequencing gel showing bands in four gel lanes that represent DNA fragments produced by the four different dideoxy sequencing reactions. The final DNA sequence is shown at the right.

a sonic pen to track locations on the X-ray film. Once the boundaries of each gel lane are identified, bases can rapidly be entered by clicking the pen on each band. However, the digitizer is only as accurate as the person using it, and it does not necessarily reduce data entry errors.

DNA sequences can also be read from autoradiography X-ray films using image analysis software. A digital image of the X-ray film must be made using a scanner. Once the shape of each lane on the gel is defined, it is possible for software to automatically identify each band and assign the correct base. Some image enhancement techniques can also be used to improve the accuracy of base calling, but usually automated base assignments are not more accurate than those made by an experienced human technician.

Applied Biosystems (ABI, Foster City, CA, USA), working with the laboratory of Lee Hood at The California Institute of Technology (6), developed and commercialized a fluorescent automated sequencer. The ABI sequencing technology uses four different fluorescent dyes incorporated into each of the four types of dideoxynucleotides. This allows the sequencing reaction to be conducted in a single tube and loaded onto a single lane of a polyacrylamide gel. Fragments ending in each of the four bases are detected fluorescently using color-specific filters. The DNA sequence is then read off of the gel automatically by a computer, eliminating the error-prone data entry step (see Figure 3). In 1995, ABI introduced machines that use capillary electrophoresis instead of tubes. More recent models have 96 capillaries and automatic sample loading. Capillary fluorescence sequencers from ABI and a few other manufacturers now account for most of the sequence data contributed to GenBank and all of the sequencing for the Human Genome Project.

SEQUENCING STRATEGIES

The biochemical techniques of sequencing only allow sequences to be determined in approximately 500-bp chunks known as reads. This is due to both the limitations of the DNA polymerase reaction and the resolution of polyacrylamide gel electrophoresis. Yet most genes contain many thousands of base pairs, and many modern sequencing projects are intended to produce complete sequences of large genomic regions or even entire chromosomes (many millions of base pairs). As a result, all sequencing projects involve the division of the target DNA into a set of overlapping approximately 500-bp fragments and the cloning of these fragments into a suitable plasmid or phage vec-

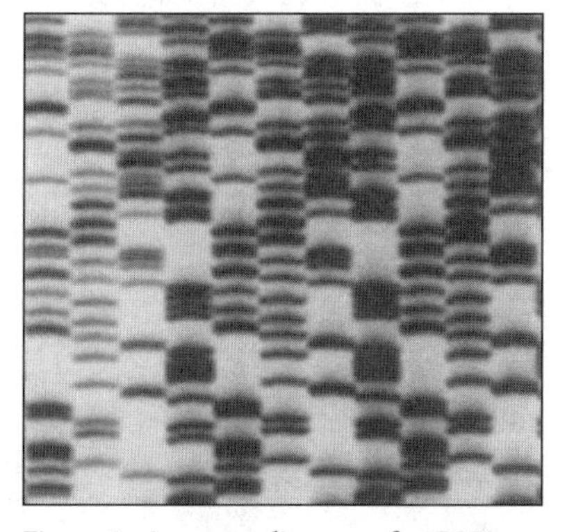

Figure 2. An autoradiogram of a DNA sequence.

tor. Once the sequences of these fragments are determined, they must be pieced back together into contigs (laboratory slang for contiguous sequenced regions) by identifying overlaps between fragments.

Since the accuracy of DNA sequencing is considerably less than 100% (both by automated sequencers and by manual methods), it is necessary to repeat the sequencing of each region of DNA several times. Since some of the inaccuracies of sequencing are caused by the specific order of bases, greater accuracy can be achieved by determining the sequence of both complementary strands of the DNA molecule. This is generally accomplished by sequencing each cloned fragment in both directions. Therefore, the final job of producing a single accurate consensus sequence requires the assembly of multiple reads of overlapping

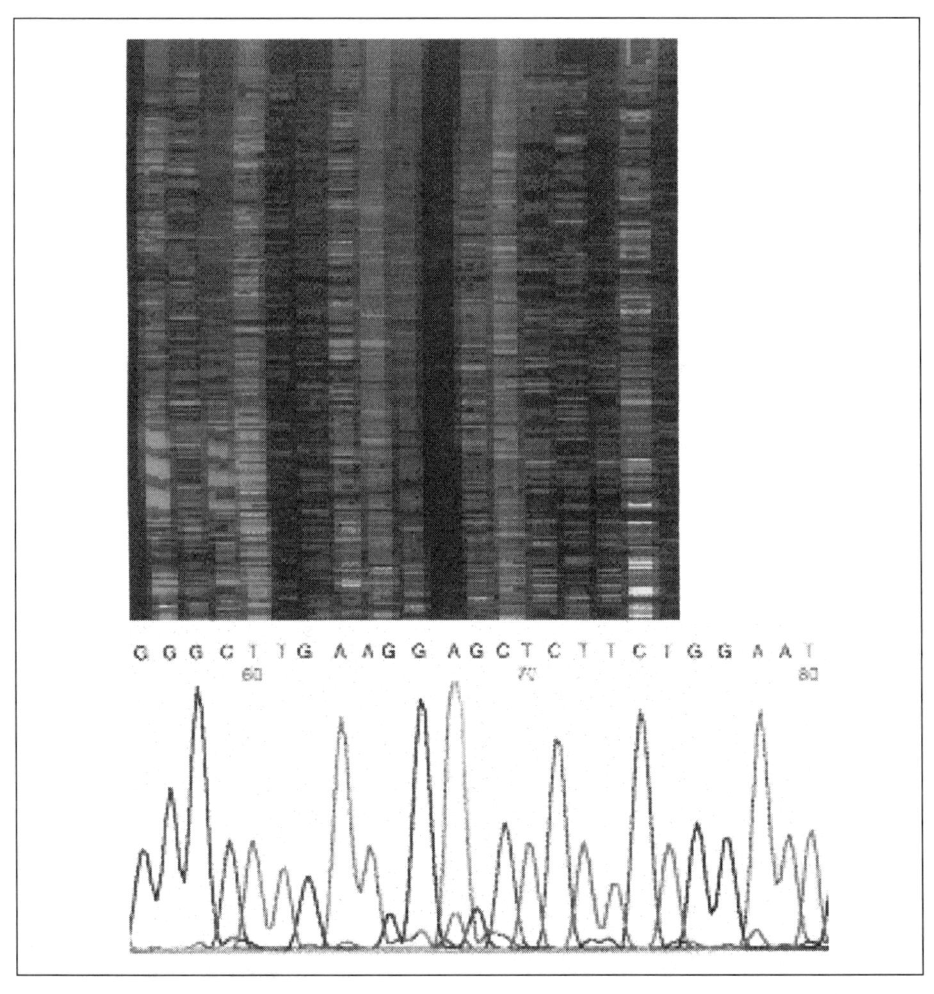

Figure 3. A portion of a fluorescent image of a DNA sequencing gel produced on an ABI automated sequencer (upper panel) and a sample chromatogram from one lane of the gel (lower panel). (See color plate A11.)

sequence fragments in both directions. Figure 4 shows a typical set of overlapping sequence reads that form a contig that could be used to accurately determine the sequence of a region of DNA. Researchers have come to rely on computers both for the identification of overlapping regions between different fragments and for the determination of the final consensus sequence.

Researchers have generally taken one of three different approaches to the planning of sequencing projects. These approaches reflect the overall strategy for dividing a large piece of DNA into 500-bp segments and then assembling the sequences of those segments into the complete sequence of the original DNA molecule.

Ordered Subfragments

The ordered subfragment strategy requires that the DNA to be sequenced is cloned into small ordered overlapping fragments, known as a nested deletion set. This strategy requires much more initial cloning work in the laboratory, usually by a process of unidirectional digestion with an exonuclease. However, it generally minimizes the number of actual sequencing reads required to complete a project and makes minimal demands of software to organize the reads because it is known how they should fit together to form the final contig.

Primer Walking

Rather than subdividing a large piece of cloned DNA, an initial approximately 500-bp read is made at each end of the target DNA fragment using primers that match sequences in the vector. Then, new primers are synthesized that match the ends of the newly determined sequences, and these primers are used to extend the sequenced region. This process is repeated with each successive sequencing reaction using information from the previous reads. Assembly problems are minimized because both the order and the amount of overlap of the reads are known. The primer walking strategy is slow since each step must wait for data from the previous step and for the synthesis of new primers.

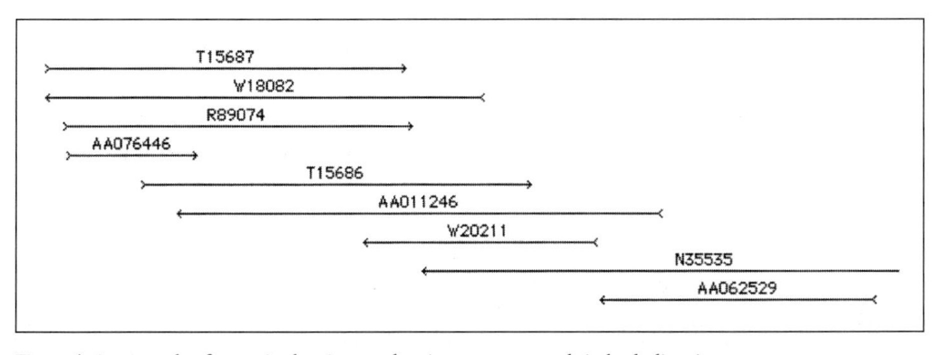

Figure 4. An example of a contig showing overlapping sequence reads in both directions.

Shotgun Sequencing

A third strategy known as shotgun sequencing (10) takes maximum advantage of the speed and low cost of automated sequencing. Many copies of the entire piece of DNA to be sequenced are randomly chopped into small approximately 500-bp fragments, which are all cloned into a plasmid vector and sequenced. This approach relies totally on software to assemble a jumble of sequence reads into a coherent and accurate contig. This approach requires many more individual sequencing reactions, but much less meticulous cloning and record keeping for a large project. The Institute for Genomic Research (TIGR) extended the power and utility of the shotgun approach by using it to determine the complete genomic sequences of microorganisms *Haemophilus influenzae*, *Methanococcus jannaschii*, *Mycoplasma genitalium*, and many others. Thus, as sequencing has become increasingly automated, it is more efficient to devote the time of skilled technicians to analyzing data after preliminary automated contig assembly by computer, rather than to careful cloning of overlapping fragments in the laboratory.

Researchers generally choose one of these three methods based on their personal perceptions of the relative cost, speed, and the size of the DNA to be sequenced. Shotgun sequencing requires the most faith in computer sequence assembly programs. Ordered subfragments are unwieldy for very large sequencing projects. Primer walking has become much more popular as the cost of custom synthesized primers has declined, but it is still slowed by the need to obtain a new sequence before the next primer is synthesized. Many sequencing projects use approaches that involve a mixture of these three basic strategies. Large sections of genomic DNA are first carefully subcloned into overlapping megabase-sized fragments [yeast artificial chromosomes (YACs) or bacterial artificial chromosomes (BACs)]. Then these large fragments are shotgun sequenced. Gaps in the assembled sequences then can be filled by primer walking.

Computer Software for Sequence Assembly

The difficult, but critically necessary, computational problem of assembly of short sequences into large contigs for sequencing projects has led to the development of a lot of different pieces of software. Each of the laboratories playing a major role in the Human Genome Project has had to deal with this problem, resulting in the development of a variety of custom mainframe computer programs. The companies that manufacture automated sequencing equipment (PE Biosystems and LI-COR [Lincoln, NE, USA]) have their own software, and most of the comprehensive molecular biology software packages have a DNA sequence assembly module.

There are several specific problems that DNA sequencing software must address. First, there is the issue of data handling and project management. This is a classic database issue and is not unique to molecular biology. Large quantities of text sequence data must be archived, and any additional information about a particular read must be saved in association with the sequence data. Each read is derived from a particular template (plasmid) and a primer (either a forward or reverse vector primer or a custom primer-walking primer). These plasmid templates are in turn derived from a subcloning operation, which might be the shotgun cloning of a cosmid, ordered deletions of a larger plasmid, etc. Sequence data that are derived from manual sequencing with radioactively-labeled DNA fragments may have computer files of scanned images of autoradiography X-ray films. Automated sequencing produces both electronic images of gels and chromatograms associated with each sequence read (Figure 3). The Macintosh® and Windows program Sequencher™ (Gene Codes, Ann Arbor, MI, USA) does a particularly good job of saving chromatogram images aligned together with the DNA sequence and using these images to aid in the sequence editing process (Figure 5), as does LaserGene™ (DNAStar, Madison, WI, USA).

Second, the software must detect and remove vector sequences from the ends of each sequence read. This can be quite challenging because the sequence at the ends of the read are the most error prone, yet the inclusion of even a single extra base of vector sequence is unacceptable in published molecular biology work. It is often necessary for a human to edit out vector sequences from contigs assembled by computers. In

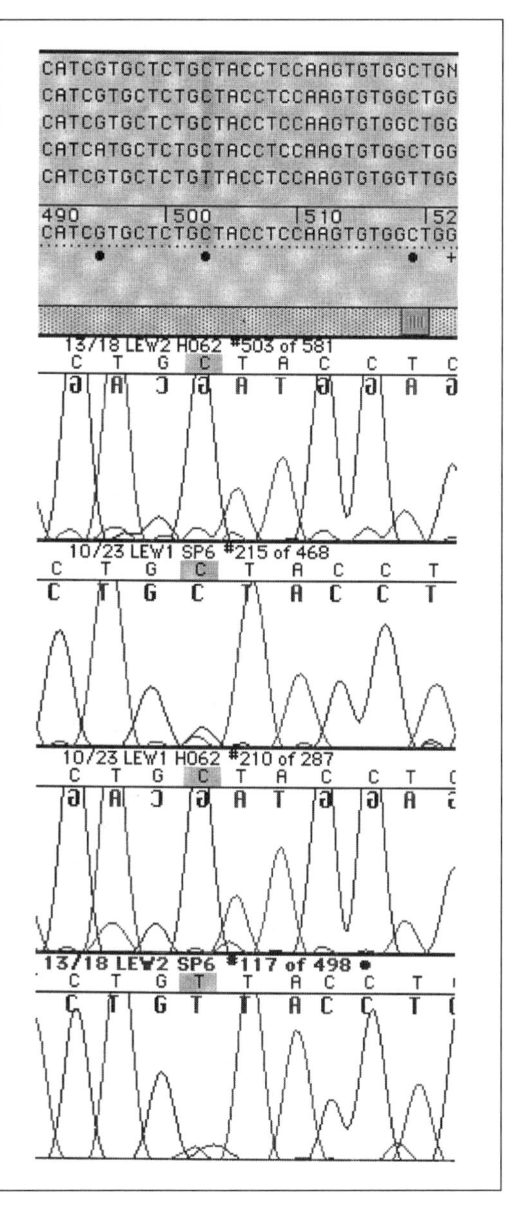

Figure 5. A sample screen from the Sequencher program showing a portion of a contig with a consensus sequence and chromatograms from several individual sequence reads. Conflicting bases are highlighted at a single position.

fact, in the laboratories that produce a large amount of sequence information, more human time is spent on final editing of assembled contigs than on the laboratory work required to produce the sequence data.

Third, the software must correctly detect overlaps between sequence reads and assemble the overlapping fragments into contigs. Most software uses some version of the approximate alignment procedure of Wilbur and Lipman (9) to detect similarity between overlapping fragments. The software must be optimized to allow only small gaps and to strongly favor regions of similarity that run off the ends of the fragments, rather than allowing matches between regions internal to the fragments (Figure 6). The software must also test the complementary strand of each sequence for similarity to all other fragments (and to their complements) because many subcloning procedures allow fragments to be cloned in either orientation. This effectively quadruples (rather than just doubles) the computational problem.

Fourth, a consensus sequence must be determined for each contig (Figure 7). Since sequencing technology is not 100% accurate, every region of DNA having its sequence determined must be sequenced several times in both directions. Then, a consensus sequence must be created that identifies the most probable base in each position. Computer programs are often unable to create an unambiguous consensus sequence from conflicting data; a human operator must scrutinize the sequences to resolve conflicts or to decide that another read is required.

The efficient use of supplemental data such as digitized images of autoradiograms or electropherograms can help a technician to make correct judgments when the multiple reads of a sequence are not in agreement. The use of this supplemental data provides a significant advantage over analysis of just the plain text of DNA sequences. This advantage means more accurate final sequences and fewer reads needed to complete a sequencing project.

Yet another problem faced by sequence alignment software is the various forms of duplication found in natural DNA sequences. DNA contains tandem repeats of single bases (AAAAAA), short patterns known as simple sequence repeats or microsatellites (GATGATGATGAT), or longer repeats known as minisatellites (generally 50–200 bp). Some entire genes, such as the ribosomal DNA genes, are repeated many times in tandem. Other repeated sequences such as transposons or the human Alu repeats occur throughout the genome. All of these repeated sequences will interfere with the operation of a sequence assembly program that just looks for similarity between the overlapping ends of approximately 500-bp reads. This problem is even more difficult in shotgun sequencing, in which the order of the reads is not known before assembly, is attempted.

SOFTWARE RECOMMENDATIONS

Sequencher™ from GeneCodes is a Macintosh and Windows program that is entirely dedicated to assisting researchers with DNA sequencing. In addition to the features common to GCG and other full-featured molecular biology packages, such as sequence entry and a multiple sequence alignment editor, Sequencher offers specialized tools for working with the output from automated sequencers. Sequencher can import the chromatograms (see Figure 7) directly from an ABI (and other manufacturers') automated sequencer. So, rather than just looking at a text file of the sequence data, the quality of the sequence at each base can be assessed, much like an autoradiogram. When multiple sequencers are aligned and mismatches are found, the electropherograms can also be aligned and decisions can be made based on the quality of the sequence rather than just which base occurs more often in multiple reads.

For those using manual sequencing with autoradiograms, Sequencher also has tools that can collect data from digitizers and can read back sequences aloud to simplify error checking. Sequencher offers impressive automated tools for rapid automatic removal of vector sequences and low-quality sequence data from the ends of reads. Other tools such as translation and restriction mapping are also available. Sequencher can handle large numbers of sequence reads and long contigs provided it is allocated enough RAM (at least 10 MB for contigs over 2000 bp). Assembly of large numbers of fragments is processor intensive—you will enjoy working with this program more on a high-speed computer.

LaserGene™ from DNAStar, Inc. (Madison, WI, USA) also contains capable and user-friendly DNA fragment assembly module called SeqMan™, which is

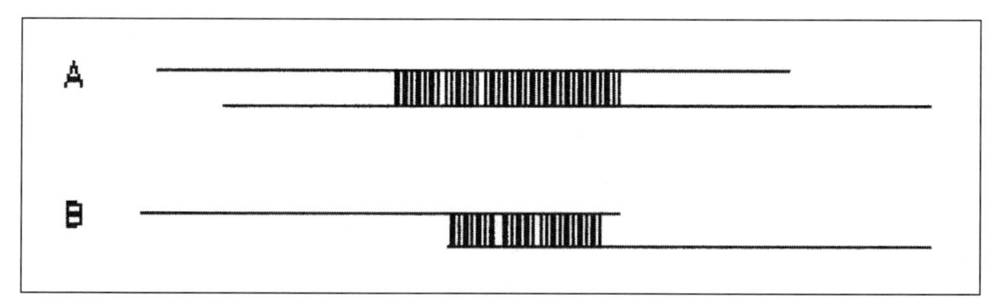

Figure 6. An example of overlapping fragments showing alignments of internal segments (A) or end segments (B).

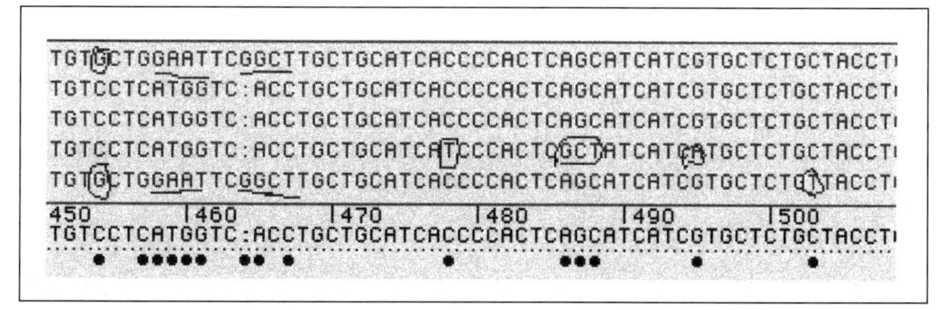

Figure 7. A portion of a contig showing a consensus sequence and individual reads that differ from the consensus. (See color plate A11.)

quite similar to Sequencher. SeqMan has automated vector and low-quality data removal and allows the user to view multiple alignments of the electropherograms from ABI (and other manufacturers') automated sequencers. SeqMan also includes the ability to evaluate sequence quality at each base across overlapping reads in order to call the most accurate consensus sequence possible. SeqMan also incorporates links to the NCBI's BLAST server for direct similarity searching against GenBank.

The **GCG** package is the most popular molecular biology software in the world today. GCG has a fragment assembly system which is a series of related programs [based on the Staden package (1)] that allow data entry and assembly of overlapping nucleotide sequence fragments into contigs. The sequence project tools include:

- **SEQED:** A single sequence editor (based on a simple mainframe text editor).
- **GELSTART:** Creates fragment assembly projects, initializes work sessions on existing projects.
- **GELENTER:** Adds individual sequences (reads) to an assembly project, allows input of new sequences from keyboard, digitizer, or import of existing text files.
- **GELMERGE:** Assembles individual sequences into contigs, can automatically remove vector sequences.
- **GELASSEMBLE:** Multiple sequence editor for viewing and editing contigs, allows manual alignment of fragments (even those that won't align with GELMERGE), insertion/deletion of gaps, and changing of individual bases.
- **GELVIEW:** Displays contigs as a schematic display of overlapping fragments.
- **GELDISASSEMBLE:** Breaks up contigs into individual sequences within a project.

These tools are relatively simple to use and can handle extremely large amounts of data (1650 fragments with a total of 380 000 bases). All data is stored as plain text files within a specified directory on a mainframe computer. These tools can only handle plain text DNA sequence data. The encoding of autoradiograms or chromatograms as text is not handled by GCG.

By default the GCG GELMERGE program looks for identical "words" that are 7 bases long between pairs of sequences and requires a minimum overlap length of 14 bases with 80% identity. When using the GELMERGE program with low-quality data (containing many incorrect or ambiguous bases), it is often necessary to use much lower values for these stringency parameters. GELMERGE can also detect and remove vector sequences with the same algorithm used for detecting overlaps between sequence fragments.

GCG released a completely new sequence assembly program in 1999 called SeqMerge™ with functionality similar to the X-windows based multiple alignment editor of SeqLab, which is in turn based on Steve Smith's GDE (Genetic Data Environment) (7). SeqMerge is also able to import and view traces (electropherograms) from ABI and other automated sequencers.

The **Staden** package (1) has continued to be developed as a very sophisticated DNA sequence assembly and analysis package for UNIX computers (a Windows version is under development). While it is substantially more complex to use than Sequencher, LaserGene, or GCG/SeqMerge, it does produce sophisticated graphical output in an X-Windows environment. The Staden package is used in many large genome sequencing centers. It is available for free for academic users, and commercial licenses are also available:
http://www.mrc-lmb.cam.ac.uk/pubseq/staden_home.html

For larger scale sequencing projects that involve the assembly of hundreds to thousands of individual gel reads, there are some more powerful (but less user-friendly) programs available. Dr. Phil Green and co-workers at the University of Washington, Seattle, have developed a set of programs known as **Phred** and

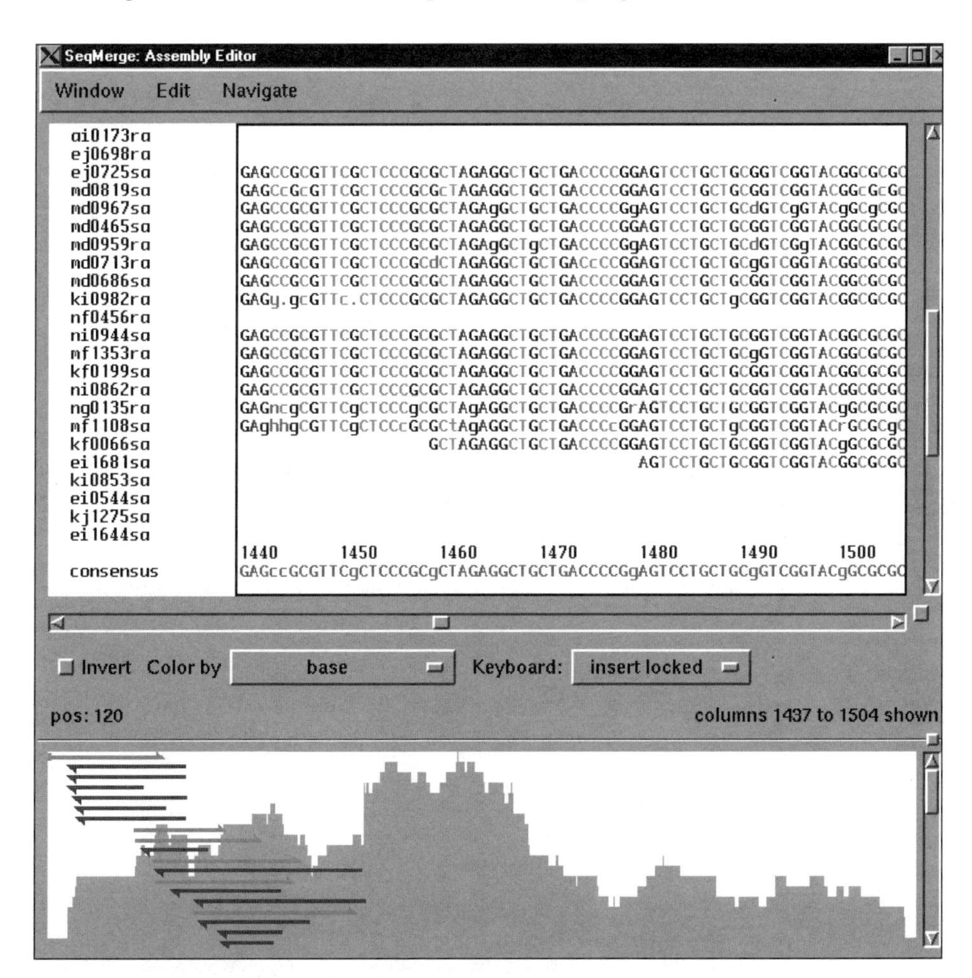

Figure 8. Sample SeqMerge screen. (See color plate A12.)

Phrap that provide automated assembly of DNA sequence trace files (2,3). Phred reads the raw data from ABI sequencing machines and calculates error probabilities for each base call. Phrap then uses these error probabilities to determine highly accurate consensus sequences by using the highest quality sequence at each position to build the consensus. These error probabilities allow researchers to concentrate their efforts on re-sequencing those areas that have the most questionable quality in the final consensus. Phred and Phrap are available for free (for non-commercial use) as UNIX-compatible source code from the author Phil Green at the University of Washington Genome Center (**www.genome.washington.edu**). A commercial version of Phred and Phrap for UNIX, Windows, and Machintosh is also available from CodonCode Corporation of Dedham, MA (**www.phrap.com**).

The Institute for Genomic Research (TIGR) has been a pioneer in large-scale sequencing projects. They completed the first full sequence of a prokaryote genome in 1995 (4) using a shotgun strategy that relied very heavily on their own sequence assembly software. Source code for the **TIGR Assembler** program (8) is available for free for non-commercial research use from TIGR (**www.tigr.org**).

REFERENCES

1. **Dear, S. and R. Staden.** 1991. A sequence assembly and editing program for efficient management of large projects. Nucleic Acids Res. *19*:3907-3911.
2. **Ewing, B., L. Hillier, M.C. Wendl, and P. Green.** 1998. Base-calling of automated sequencer traces using Phred. I. Accuracy assessment. Genome Res. *8*:175-185.
3. **Ewing, B. and P. Green.** 1998. Base-calling of automated sequencer traces using Phred. II. Error probabilities. Genome Res. *8*:186-194.
4. **Fleischmann, R.D., M.D. Adams, O. White, R.A. Clayton, E.F. Kirkness, A.R. Kerlavage, C.J. Bult, J.F. Tomb, B.A. Dougherty, J.M. Merrick et al.** 1995. Whole-genome random sequencing and assembly of Haemophilus influenzae Rd. Science *269*:496-512.
5. **Sanger, F., S. Nicklen, and A.R. Coulson.** 1977. DNA sequencing with chain-terminating inhibitors. Proc. Natl. Acad. Sci. USA *74*:5463-5467.
6. **Smith, L.M., J.Z. Sanders, R.J. Kaiser, P. Hughes, C. Dodd, C.R. Connell, C. Heiner, S.B.H. Kent, and L.E. Hood.** 1986. Fluorescence detection in automated DNA sequence analysis. Nature *321*:674-679.
7. **Smith, S.W., R. Overbeek, C.R. Woese, W. Gilbert, and P.M. Gillevet.** 1994. The genetic data environment an expandable GUI for multiple sequence analysis. Comput. Appl. Biosci. *10*:671-675.
8. **Sutton, G., O. White, M. Adams, and A. Kerlavage.** 1995. TIGR Assembler: a new tool for assembling large shotgun sequencing projects. Genome Sci. Tech. *1*:9-19.
9. **Wilbur, W.J. and D.J. Lipman.** 1984. The context dependent comparison of biological sequences. SIAM J. Appl. Math. *44*:557-567.
10. **Wilson, R., R. Ainscough, K. Anderson, C. Baynes, M. Berks, J. Bonfield, J. Burton, M. Connell, T. Copsey, J. Cooper et al.** 1994. 2.2 Mb of contiguous nucleotide sequence from chromosome III of C. elegans. Nature *368*:32-38.

10 Phylogenetics

TAXONOMY AND PHYLOGENETICS

The theory of evolution is the foundation upon which all of modern biology is built. From anatomy to behavior to genomics, the scientific method requires an appreciation of changes in organisms over time. It is impossible to evaluate relationships among gene sequences without taking into consideration the way these sequences have been modified over time.

Similarity searches and multiple alignments of sequences naturally lead to the question: "How are these sequences related?" and more generally "How are the organisms from which these sequences come related?" After working with sequences for a while, one develops an intuitive understanding that for a given gene, closely related organisms have similar sequences and more distantly related organisms have more dissimilar sequences. Also, it seems logical that given a set of similar sequences, it should be possible to reconstruct the evolutionary relationships (ancestral relationships) among genes and among organisms. This involves creating a branching structure, termed a phylogeny or tree, that illustrates the relationships by descent among the sequences. This is a graph composed of nodes, branches, and endpoints at the tips of the branches as show in Figure 1. The branch tips represent either taxonomic units such as individual species (a species tree) or individual sequences (a gene tree). The internal nodes represent branching points of evolution where two subgroups diverged from a common ancestral population. The branch lengths are generally meant to indicate relative evolutionary distances between the tips and nodes, but some trees are created with nonproportional branch lengths. Some trees, such as the one show in Figure 1, have all of their branches emerging from a common node know as the "root", which is meant to represent a single common ancestor. An unrooted tree leaves the position of the common ancestor unspecified.

Underlying these intuitive understandings about the relationships between sequences are a set of assumptions about the nature of evolution at a molecular level. These concepts were first developed by Linus Pauling and Emile Zuckerkandl in 1965 (25).

Bioinformatics
By Stuart M. Brown

First, assume that mutations occur more or less randomly in DNA and therefore in protein sequences. Second, these mutations accumulate with time. Therefore, the number of differences in homologous protein or DNA sequences between any pair of organisms is proportional to the time since those organisms diverged from a common ancestor.

If similar proteins in closely related species were derived from a single protein in the common ancestor of those two species, then all of the diverse proteins in all of the organisms in the biosphere must trace their ancestry back to a single set of proteins in the original ancestor of all life on Earth. There are a number of different processes that cause the divergence of protein sequences.

1. Genetic drift: identical genes in different species experience different mutations and are subject to different selection pressures.
2. Gene duplication: single genes (or whole chromosomes) can be duplicated, then duplicated genes are free to mutate and take on new functions without the loss of original functions.
3. Recombination: translocations, insertions and deletions, transposons, retroviruses, etc., can mix up chunks of genes into new combinations. Splicing seems to be especially common at intron/exon borders. Some chunks of

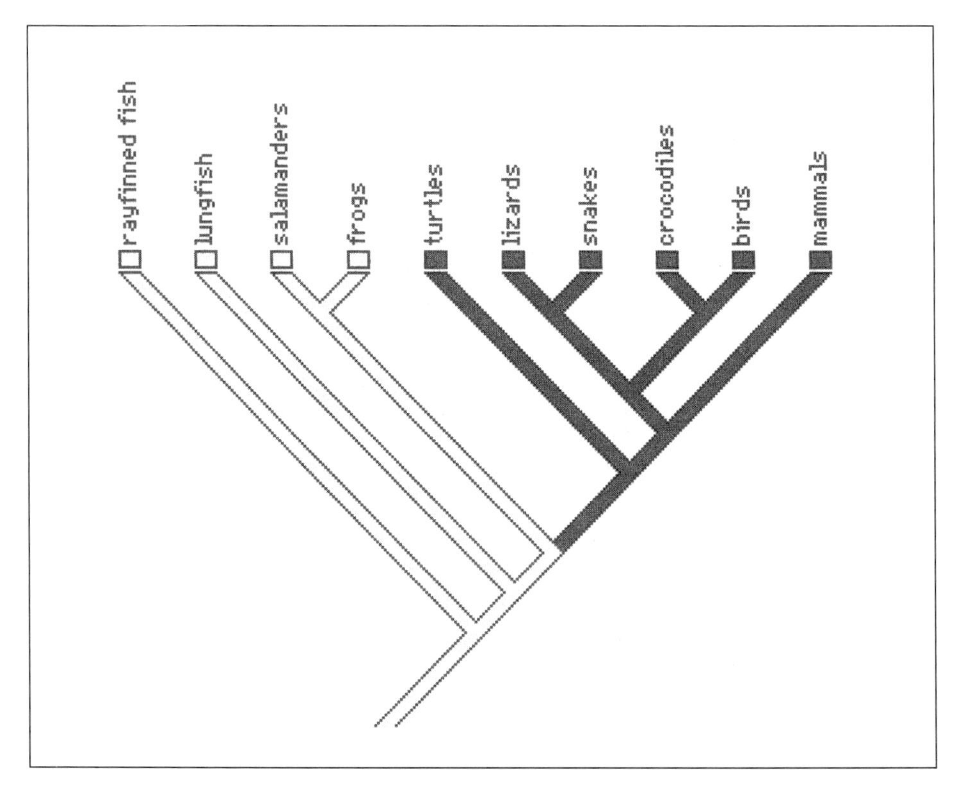

Figure 1. Sample phylogenetic tree for vertebrates (see Reference 16).

DNA are functional units that have been reused in a wide variety of genes (i.e., exon shuffling).

The study of the relationships between groups of organisms is called taxonomy, an ancient and venerable branch of classical biology that dates back at least to the time of the classical Greek philosophers. Taxonomy was formalized as a scientific discipline know as systematics by Carolus Linnaeus (1707-1778). The branch of taxonomy that deals with numerical data such as DNA sequence is known as phylogenetics. This subject also overlaps significantly with a branch of evolutionary biology known as molecular evolution and with the mathematical and computer-intensive discipline of population biology.

Taxonomy has the mystique of an old discipline, practiced in musty rooms in the subbasement of museums where scientists spend all of their time staring through magnifying glasses at jars of pickled specimens. However, taxonomists were early adopters of the new technologies of molecular biology because DNA sequences have many advantages over more classical types of visible taxonomic characters.

1. Character states can be scored unambiguously (genotypes can be read directly rather than relying on phenotype).
2. Huge numbers of characters can be scored for each individual.
3. Information can be obtained on both the extent and the nature of divergence between sequences (nucleotide substitutions, insertion/deletions, or genome rearrangements).
4. Comparisons can be made between groups of organisms (populations) with minimal phenotypic differences.
5. Phylogenies can be built for groups of organisms that are so widely diverged that they do not share many morphological traits.

DNA and protein sequences can be considered as huge collections of phenotypic characters. Each DNA base or amino acid can be an individual phenotypic character, or just those that differ among a group of sequences (i.e., informative characters). Clusters of DNA bases can be indirectly scored as characters (genetic markers) by the use of restriction enzymes to evaluate restriction fragment-length polymorphisms (RFLPs) or by the use of the polymerase chain reaction (PCR). All of these methods (as well as many other variants) lead to the collection of data from the genomes of organisms.

However, there are some problems inherent in the use of DNA sequences as phylogenetic characters. Current DNA sequencing and computer technology do not allow taxonomists to compare the sequences of the entire genomes for all of the organisms that they are interested in classifying. Therefore, a great deal of molecular biology must be considered when making the choice of which sequences will be compared.

First, it is necessary to compare homologous sequences. Evolutionary information can only be obtained from the measurement of differences between sequences that share a common ancestor. The choice of which type of sequences to use for data is determined primarily by the evolutionary distance between the organisms that will be compared. Contrary to the hypothesis of Zuckerkandl and Pauling, different regions of DNA accumulate mutations at strikingly different rates. Closely related organisms must be compared using highly variable sequences, and highly diverged organisms must be compared using sequences that have been conserved over evolutionary time. Furthermore, mutation rates (and therefore the rate of evolution) may be different in different types of organisms or even within different lineages of related organisms.

DNA which codes for protein is under much stricter evolutionary control than noncoding DNA. Any change in the sequence of coding DNA (a mutation) could create a change in its protein product and possibly a change in the phenotype of the organism. Due to the redundancy of the genetic code, there are some mutations in DNA sequence, usually at the third position of a codon, that do not change the corresponding amino acid. Since random changes in protein sequences may be deleterious to the organism, there is strong selection pressure against changes in coding DNA. Some proteins can tolerate many changes without significant loss of function. For example, fibronectin is a structural peptide that shows high levels of sequence divergence among closely related organisms. Other proteins have structures that cannot be changed even a small amount without substantial loss of function and dire phenotypic effects; for example, histone H3 is 96% identical between cow and pea. The neutral theory of evolution [promoted by Kimura (14)] argues that most mutations observed in DNA and protein sequences (for example, isozyme alleles or RFLPs) have no effect on the fitness of organisms and are thus passed on to progeny strictly according to the rules of chance genetic segregation.

Mutations in noncoding gene promoters and other regulatory regions may have powerful phenotypic effects caused by changes in gene expression. Introns do not code for protein, but a mutation that affects RNA splicing can destroy the functional protein. Most other noncoding DNA is not under such tight evolutionary restraints, but there are some additional considerations, such as histone binding sites and origins of DNA replication, that may constrain mutations.

The distinction between coding and noncoding DNA sequences is relatively clear. A simple functional approach would be to define coding DNA as that found in cDNA sequences. However, this distinction ignores the important genetic process of gene duplication. Duplicate copies of genes are often freed of many evolutionary constraints—one copy of the gene can continue to make functional protein while the other is free to mutate. Gene duplication leads to the formation of gene families—which can greatly complicate the process of molecular taxonomy with questions of orthologous versus paralogous copies of similar genes. Did a gene duplication event occur before or after two species diverged from a common ancestor? Do the two copies of a gene in one species diverge more rapidly than two homologous genes in two related species? Is the copy of the gene being studied in organism A directly analogous to the copy being studied in organism B, or do they have different functions? Also, some duplicate copies of genes can become inactivated into pseudogenes, which are no longer expressed as protein, and are therefore free to accumulate mutations without selection. Any comparison of functional genes with a pseudogene will lead to inaccurate phylogenies.

The use of DNA sequences as taxonomic characters generates huge amounts of data, which require computers for analysis. There were many computer programs created for the analysis of taxonomic data long before DNA sequence data became available. However, many of these programs rely on theoretical assumptions that are not valid for DNA data. New programs created explicitly to deal with DNA data must still exist within the complex framework of theoretical population biology and phylogenetics.

SOME CAVEATS

Before describing any theoretical or practical aspects of phylogenetics, it is necessary to give some disclaimers. This area of computational biology is an intellectual minefield. Many different methods for calculating phylogenetic trees from sequence data have been developed, but neither the theory nor the practical applications of any of these algorithms are universally accepted throughout the scientific community (21). The application of different software packages to a data set is very likely to give different answers, and minor changes to a data set are also likely to profoundly change the result.

Despite all of these caveats, it is possible to calculate phylogenetic trees for data sets. Provided the data are clean, outgroups are correctly specified, appropriate algorithms are chosen, no assumptions are violated, etc., can the true, correct evolutionary tree be found and proven to be scientifically valid for any given collection of groups of organisms? Unfortunately, it is impossible to ever conclusively state what is the true tree for a group of sequences (or a group of organisms); taxonomy is constantly under revision as new data is gathered. At best, taxonomy is a science that is constantly making more and more precise guesses about the evolutionary history of organisms and their genomes. It is impossible to physically observe the process of evolution over millions of years. Instead, evolution must be inferred by comparing organisms that are alive today.

It is necessary to consider the results of phylogenetic analysis conducted with molecular data in the context of the full body of taxonomic and biological knowledge. If calculated correctly, relationships derived from sequence data actually represent the relationships between genes, which are not necessarily the same as the relationships between whole organisms. The sequences that were used for data may not have had the same phylogenetic history as the species within which they were contained. Different genes (and different species) evolve at different speeds, and there is always the possibility of horizontal gene transfer either by hybridization, vector-mediated DNA movement, or direct uptake of DNA.

CLADISTIC VERSUS PHENETIC ANALYSIS METHODS

Within the field of numerical taxonomy, there are two different methods and philosophies of building phylogenetic trees. One method is known as the phenetic approach. In this approach, a tree is constructed by considering the overall phenotypic similarities of a set of species without trying to understand the evolutionary history that has brought the species to their current phenotypes. Large amounts of data are collected that describe the states of various characters. All characters are treated as equally important. This data is then used to classify organisms into trees based on the absolute number of characters that they share. Since a tree constructed by this current-data-only method does not necessarily reflect evolutionary relationships, but rather is designed to represent phenotypic similarity, trees constructed via this method are called phenograms. A phylogenetic tree per se is often termed a dendrogram (a branching order that may or may not be the correct phylogeny). Computer algorithms based on the phenetic model generally rely on distance methods for the calculation of relationships and building of trees.

The second approach to phylogenetics is called cladistics. Cladistics was developed by Willi Hennig, a German entomologist, in the 1950s, but did not become well known until his work was translated into English in the 1960s (9,10). Cladistics is the study of clades, which are groups of related organisms. Thanks to Linnaeus, all of biology is divided into a hierarchical system from the largest classifications such as plant versus animal (kingdom) down to the level of individual genus and species. Via cladistic methods, these hierarchical groupings are viewed as a tree that is reconstructed by considering the various possible pathways of evolution (branching points between groups of ancestral organisms) and choosing from among these the best possible tree. Known ancestral relationships are considered as well as current data. Instead of looking at all the phenotypic characters of organisms, cladistics uses only a subset of characters that are uniquely shared by a group of organisms and not present in their ancestors. Trees reconstructed via these methods are called cladograms. Computer algorithms based on the cladistic model generally rely on parsimony or maximum likelihood methods for the calculation of relationships and building of trees.

For character data (physical traits of organisms such as morphology of organs, etc.) and for higher (or perhaps we should say deeper) levels of taxonomy, the cladistic approach is almost certainly superior. However, cladistic methods are often difficult to implement, requiring assumptions that are not always satisfied with molecular data and in-depth knowledge about the biology of the organisms being studied. Molecular data make good pure phenotypic characters because there is no inherent reason for a researcher to think that one mutation is more important or more ancestral than another. Phenetic approaches lead to generally faster algorithms, and they often have nicer statistical properties for molecular data.

In some cases your choice of phylogenetic algorithm may be limited by the software that is accessible, which may in turn be limited by the computing hardware that is present in your laboratory. If this is the case, then it is just as important to know the limitations of the algorithms that you use, just as it would be if you had to select from a wide variety of different algorithms.

CALCULATING DISTANCES

It is often useful to measure the genetic distance between two species, between two populations, or even between two individuals. For example, if you have two individuals who come to a hospital, and they both have the same genetic disease, you might want to know if they are related and if they might therefore have inherited the same gene. Otherwise, this might be a manifestation of two separate mutations. Distance-based phylogeny methods require the calculation of pairwise distances between all sequences that will be used to build the tree—thus creating a distance matrix.

DNA Distances

Distances between DNA sequences are relatively simple to compute as the sum of the single-base differences between two sequences. Distances can only be calculated for pairs of homologous sequences that are similar enough to be aligned. This method works quite well for simple point mutations (changes from one base to another) and small insertions/deletions or indels (which are considered as gaps in the alignment), but cannot provide an accurate measure for large insertions, deletions, and recombinations. These large-scale DNA rearrangements are handled better by cladistic methods as a single character with an appropriate weight rather than blindly adding up the large number of resulting single-base changes in the alignments.

Distance methods give a single measure of the amount of evolutionary change between two genomes since divergence from a common ancestor. Either all base changes are considered equally or a simple matrix of the frequencies of the 12 possible types of replacements (each base can be replaced by one of the three other bases) can be used. Differences due to indels are generally given a larger weight than replacements, but indels of multiple bases at one position are given less weight than multiple independent indels.

Protein Distances

Distances between amino acid sequences are a bit more complicated to calculate. From a functional standpoint, some amino acids can replace one another with relatively little effect on the structure and function of the final protein, while other replacements can be devastating. From the standpoint of the genetic code, some amino acid changes can be made due to the replacement of a single DNA base, while others require two or even three changes in the DNA sequence.

In practice, what has been done is to calculate tables of frequencies, such as the PAM (percent accepted mutations) matrix (see Chapter 5 entitled Similarity Searching), of all amino acid replacements between sets of related amino acid sequences (protein families) in the databanks (3). Distances between two aligned protein sequences are calculated by application of the PAM matrix (or another scoring matrix) to each pair of amino acids to generate a distance measurement for that position and then summing the distances over the entire alignment.

One problem that plagues all distance methods is multiple substitutions at a single site, which is more common in distant relationships and for rapidly evolving sites. For nucleic acid sequences, the most popular methods to correct for multiple substitutions are Kimura's two-parameter method (14), the Tajima-Nei method (22), and the Jin-Nei gamma distance method (12). For protein sequences, the Kimura method is generally used; and for either type of sequence, the Jukes-Cantor method (13) is used. Distance measurements also allow for some measurement of the reliability of the final tree by the calculation of a variance which is computed from the variances of each entry in the initial distance matrix.

CLUSTERING ALGORITHMS

Once all of the pairwise distances have been calculated among a set of sequences, those sequences can be grouped into a tree. There are a variety of different clustering algorithms that can be used to build trees from distance data. Cluster algorithms can be applied to many different types of molecular data including isozymes, restriction sites, RFLPs, etc., but we will restrict this discussion to DNA and protein sequence data.

Unweighted Pair Group Method Using Arithmetic Averages

The simplest of the distance methods is a type of cluster algorithm that is known as UPGMA (unweighted pair group method using arithmetic averages) (19). This method has gained popularity mostly because of its simplicity and also because of its speed (though many other distance methods are as fast).

UPGMA takes raw pairwise distance data among a set of sequences in the form of a matrix and uses these distances to group first the least distant pair, then adds the next, etc., until all sequences are grouped. No corrections or additional computations are applied to the data. Clearly this lack of processing is appealing to the pure "data as we find it" philosophy of phenetic analysis, but it generally does not reflect the reality of molecular evolutionary processes. In particular, UPGMA ignores the tendency for multiple mutations to occur at one site, thus underestimating relative distances between more distantly related sequences.

The popular Genetics Computer Group (GCG) multiple alignment program PILEUP uses UPGMA to create its initial dendrogram of DNA sequences. It then uses this dendrogram to guide its multiple alignment algorithm, first aligning the most similar sequences in pairs, then adding individual sequences or other aligned clusters until all sequences are included in the alignment.

Neighbor Joining

Another very popular distance method is the neighbor joining method (17). This method attempts to correct the UPGMA method for its (frequently invalid) assumption that the same rate of evolution applies to each branch. Hence, this method yields an unrooted tree. A modified distance matrix is constructed to adjust for these differences in the rate of evolution of each taxon. Similar to the UPGMA method, the least distant pairs of nodes are linked, their common ancestral node is added to the tree, and their terminal nodes are pruned from the tree. This continues until only two nodes remain.

Others

There are a wide variety of other distance-based clustering algorithms constructed with differing sets of assumptions. Neighbor joining has given the best results in simulation studies and it is the most computationally efficient of the distance algorithms (18). However, the choice of algorithm must reflect the type of data being analyzed and the history of the subdiscipline from which the data is derived (i.e., immunology versus numerical plant taxonomy).

CLADISTIC METHODS: PARSIMONY

Cladistic methods of molecular phylogenetic analysis are based on the explicit assumption that a set of sequences evolved from a common ancestor by a process of mutation and selection without mixing (hybridization or other horizontal gene transfers). Many of these methods work best if a specific tree is already known, so that statistics can be compared between a finite number of alternate trees rather than calculating all possible trees for a given set of sequences.

Parsimony is the most popular method for reconstructing ancestral relationships (21). Parsimony allows the use of known evolutionary information in tree building (in contrast to distance methods, which lose much data by compressing all differences between sequences into a single number). This method involves evaluating all possible trees (in practice, usually only a subset is examined) and giving each a score based on the number of evolutionary changes (mutations) that are needed to explain the observed data. The most parsimonious tree is the one that requires the fewest evolutionary changes for all sequences to derive from a common ancestor (4). This is easiest to explain by example. Consider four sequences: ATCG, TTCG, ATCC, and TCCG. Imagine a tree that branches once at the first position, thus grouping ATCG and ATCC on one branch, and

TTCG and TCCG on the other branch. Then, each branch subdivides into two subbranches for a total of 3 nodes in the tree (Figure 2A). Counting backward from the bottom, each sequence is separated from the root by two nodes, so the sum of the changes is equal to 8. This is a more parsimonious tree than one that first divides ATCC on its own branch, then splits off ATCG, and finally divides TTCG from TCCG (Figure 2B). This tree also has three nodes, but when all of the distances back to the root are summed, the total is equal to 9. The best (i.e., most parsimonious) tree is always the one with the smallest total branch length.

As the sequences being investigated get longer and the number of taxa increase, the number of possible trees increases very dramatically (the number of trees increases linearly with increased sequence length but as the square of the number of taxa). With real experimental data sets, it is generally not possible, even with powerful computers, to evaluate every possible tree (known as an exhaustive or brute force search) in order to find the one with the fewest evolutionary changes. Computer software must make use of heuristic shortcuts (e.g., approximations and limiting the total number of possible tree topologies to be searched). Heuristic methods begin with an approximate tree, generally calculated by neighbor joining or a similar stepwise method. Then, various branches and subtrees of this initial tree are rearranged according to some algorithm to see if this improves the score. If a better tree is found, it is used as the basis for additional rounds of rearrangement. Heuristic methods greatly reduce the computing time required to find near-optimal trees, but leave open the possibility that some better, but completely different, tree topology has been overlooked.

The branch-and-bound method (8) is a compromise between the faster heuristics and the more thorough exhaustive methods. All possible trees are built by adding nodes one at a time, but the scores of the partial tree are computed

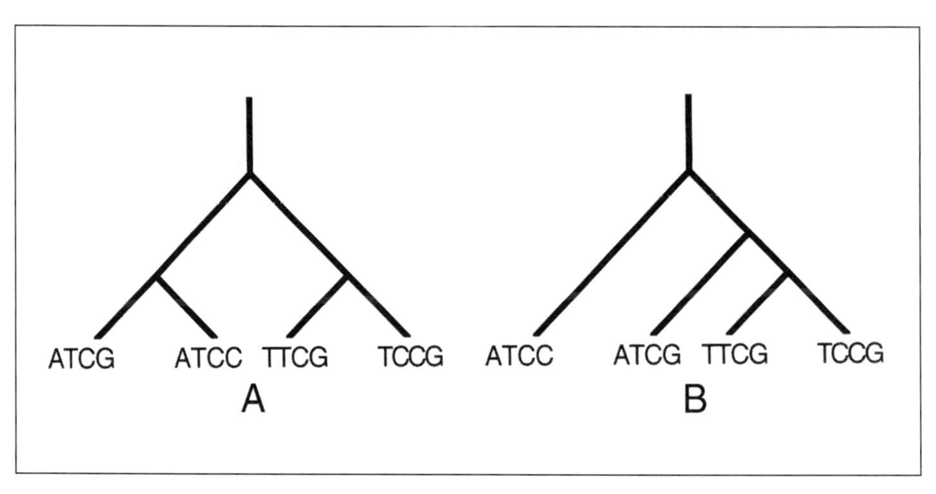

Figure 2. Parsimony method of reconstructing ancestral relationships showing 2 trees with 3 nodes each, but the sum of changes back to the root are 8 (A) and 9 (B).

each time a branch is added. If the score of the partial tree is worse than that of the best tree found so far, then this set of trees is abandoned, and a different partial tree is used as the next starting point for adding branches. A branch-and-bound search is somewhat faster than an exhaustive search, but it is also not guaranteed to find the optimal tree(s). Branch-and-bound searches are not practical with very large data sets.

In evaluating a tree built with parsimony, it is important to look at many of the near-optimal trees as well as the single best one. In some cases, trees with radically different topologies will have nearly the same total branch lengths. This is not a stable situation because slight changes in the data or in the computation methods can lead to different results (7).

CLADISTIC METHODS: MAXIMUM LIKELIHOOD

The method of maximum likelihood attempts to reconstruct a phylogeny using an explicit model of evolution. In other words, what sequence of evolutionary events was most likely to produce the current data set. Certainly, for this given model of evolution, no other method will perform as well nor provide you with as much information about the tree. Unfortunately, this is computationally difficult to do, and hence, the model of evolution must be a simple one. Even with simple models of evolutionary change, the computational task is enormous, and this is the slowest of all methods.

Given an evolutionary model, an explicit statement can be made about the probability of change from one nucleotide to another at each position in a set of aligned sequences within a specified time period. Since each nucleotide site evolves independently, the phylogeny can be calculated separately for each site. The product of the likelihoods for each site provides the overall likelihood of the observed data. To maximize the likelihood, different values are analyzed until a set of branch lengths/mutation rates are found, which provide the highest likelihood of observing the actual sequences. Finally, many different tree topologies are searched to find the best one (11). Likelihood methods generally perform better than both parsimony and distance methods, but they are not widely used due to the tremendous computational resources required to analyze real data sets.

COMPUTER SOFTWARE

GCG offers phylogenetic programs including both a set of tools for phenetic (distance-based) clustering and the complete phylogenetic analysis using parsimony (PAUP)/maximum likelihood package. The free phylogenetics inference package (PHYLIP) includes tools for distance, parsimony, and maximum likelihood analysis (6). PAUP is also available as a stand-alone package (20). The shareware program MacClade is a useful tool for visualizing and optimizing trees created using other phylogenetic programs (16).

GCG

The simple GCG program DISTANCES calculates pairwise distances between a group of sequences and writes the output into a matrix file that can then be used by the program GROWTREE to draw a tree based on either UPGMA or the neighbor joining method. Absolute distance calculations can be corrected for multiple substitutions at a site.

The entire PAUP program (20) was incorporated wholesale into GCG version 9.1 and released in November 1997. All program functions of PAUP have been divided among two GCG programs, PAUPSEARCH, which calculates trees, and PAUPDISPLAY, which produces graphical versions of PAUPSEARCH tree files. PAUPSEARCH is the single most complex GCG program with approximately 80 different optional command line parameters, some of which have many settings.

GCG's PAUPSEARCH program takes a group of aligned sequences in GCG format directly as input without having to convert them first into PAUP's native NEXUS format, a significant advantage over the stand-alone PAUP program.

PAUP

PAUP was created by David Swofford (20) and is the copyrighted property of the Smithsonian Institution. It is now distributed commercially by Sinauer Associates, Sunderland, MA, USA.

Starting with a set of aligned sequences, PAUP can search for phylogenetic trees that are optimal according to parsimony, distance, or maximum likelihood criteria using heuristic, branch-and-bound, or exhaustive tree searching methods, reconstruct a neighbor joining tree, or perform a bootstrap analysis. PAUP requires four or more aligned sequences as input.

PAUP can consume huge amounts of computer time. The exhaustive or branch-and-bound searches simply cannot be done for more than about 10 sequences of moderate length. Maximum likelihood analysis requires at least 60 times more computations than parsimony and distance methods.

PHYLIP

PHYLIP, developed by Joseph Felsenstein (6), is a free package of programs for inferring phylogenies and carrying out certain related tasks. The package is available for all major computer platforms (Macintosh®, DOS, Windows®, UNIX®, VMS®, etc.). At present, it contains 31 programs, which carry out different algorithms on different kinds of data. The programs in the package are:

1. Programs for molecular sequence data.
   - PROTPARS: protein parsimony.
   - DNAPARS: parsimony method for DNA.
   - DNAMOVE: interactive DNA parsimony.

- DNAPENNY: branch-and-bound for DNA.
- DNACOMP: compatibility for DNA.
- DNAINVAR: phylogenetic invariants.
- DNAML: maximum likelihood method.
- DNAMLK: DNAML with molecular clock.
- DNADIST: distances from sequences.
- PROTDIST: distances from proteins.
- RESTML: ML for restriction sites.
- SEQBOOT: bootstraps sequence data sets.
- COALLIKE: coalescent likelihoods from sampled phylogeny estimates.

2. Programs for distance matrix data.
   - FITCH: Fitch-Margoliash and least-squares methods.
   - KITSCH: Fitch-Margoliash and least-squares methods with evolutionary clock.
   - NEIGHBOR: neighbor joining and UPGMA methods.
3. Programs for gene frequencies and continuous characters.
   - CONTML: maximum likelihood method.
   - GENDIST: computes genetic distances.
   - CONTRAST: computes contrasts and correlations for comparative method studies.
4. Programs for discrete state data (0/1).
   - MIX: Wagner, Camin-Sokal, and mixed parsimony criteria.
   - MOVE: Interactive Wagner, Camin-Sokal, and mixed parsimony program.
   - PENNY: finds all most parsimonious trees by branch-and-bound.
   - DOLMOVE, DOLPENNY: same as preceding three programs, using the Dollo and polymorphism parsimony criteria.
   - CLIQUE: compatibility method.
   - FACTOR: recodes multistate characters.
5. Programs for plotting trees and consensus trees.
   - DRAWGRAM: draws cladograms and phenograms on screens, plotters, and printers.
   - DRAWTREE: draws unrooted phylogenies on screens, plotters, and printers.
   - CONSENSE: majority-rule and strict consensus trees.
   - RETREE: reroots, changes names and branch lengths, and flips trees.

The use of PHYLIP is really a subject for an entire book—or a multivolume series—so it is not possible to outline the program functions here. Suffice it to say that all programs work with a command line interface, and input data formats are extremely finicky, but once it is fed the data, most computations work quite rapidly on even very modestly powered desktop computers. Output for most programs is strictly ASCII text, and it is generally an agonizing process to create graphical versions of phylogenetic trees.

MACCLADE

MacClade provides a graphical and interactive analysis of phylogeny and character evolution (16). It also provides many tools for entering and editing data and phylogenies and for producing tree diagrams and charts. The user can manipulate cladograms on screen as MacClade gives diagnostic feedback such as tree length, and the results are depicted in graphics and charts. Tools are provided to move branches, reroot clades, create polytomies, and automatically search for more parsimonious trees.

MacClade works well in conjunction with PAUP to view and edit trees and to create high-quality graphical output. MacClade allows both experienced systematists and novices to visually explore phylogenetic trees and observe the effects of modifications to these trees.

MacClade is a Macintosh program that was created by Wayne Maddison and David Maddison, now distributed by Sinauer Associates for $100. An excellent demo version is available on the compact disc that accompanies this book.

PHYLOGENETICS ON THE WEB

There are a few computing centers that have recently begun to offer free phylogenetics servers on the Web. This is quite brave of them considering the potential for these programs to consume huge amounts of computing power if large numbers of long sequences are used. The amount of computation is also dependent on the type of algorithm used: distance methods (UPGMA and neighbor joining) are quick, parsimony methods take longer, and maximum likelihood methods are very slow.

The Institut Pasteur, Paris has a PHYLIP server at (**http://bioweb.pasteur.fr/seqanal/phylogeny/phylip-uk.html**).

Joaquin Dopazo at the Centro Nacional de Biotecnologia, Madrid, Spain has made a very nice server interface for PHYLIP (**http://www.cnb.uam.es/cgi-bin/dopazo/PHYLIP/phylip**).

The Belozersky Institute at Moscow State University in Russia has their own GeneBee phylogenetics server (1,2) (**http://www.genebee.msu.su/services/phtree_reduced.html**).

Another challenge in phylogenetic analyses is producing a publication-quality printout of your results. The PHYLIP drawtree program is rather poor in this regard, both in terms of user interface and in the quality of the final product. The Phylodendron Web site (**http://iubio.bio.indiana.edu/treeapp/treeprint-form.html**) is a tree drawing program with a better user interface and a lot of formatting options including PDF (Portable Document Format, Adobe Systems, Seattle, WA, USA). Whatever program you use, you may end up taking the final tree into a graphics program and redrawing it.

There is a wealth of basic information about taxonomy and molecular evolution available on the Web.

For general background reading, explore the University of California Museum of Paleontology's Journey Into Phylogenetic Systematics (**http://www.ucmp.berkeley.edu/clad/clad4.html**).

The Tree of Life Project at the University of Arizona gives a good overview of the overall process of taxonomy (**http://phylogeny.arizona.edu/tree/phylogeny.html**).

Joe Felsenstein, the author of PHYLIP, maintains an extensive Web site that offers free downloads of PHYLIP software for all computing platforms, PHYLIP documentation and FAQs, and information about 160 other phylogeny software packages (**http://evolution.genetics.washington.edu/phylip.html**).

The Willi Hennig Society maintains a Web site that contains an index of all articles from the journal Cladistics, a searchable database of phylogenetics literature, and an extensive list of phylogenetics software (**http://www.vims.edu/~mes/hennig/hennig2.html**).

The Society of Australian Systematic Biologists have posted an excellent review article titled *Introduction to Phylogenetic Systematics* by Peter H. Weston and Michael D. Crisp (23), and *Introduction to Some Computer Programs Used in Phylogenetics* by Michael D. Crisp (**http://www.science.uts.edu.au/sasb/WestonCrisp.html**) and (**http://www.science.uts.edu.au/sasb/programs.html**).

An excellent introductory manual to phylogenetic analysis (using cladistics) by Diana Lipscomb of George Washington University (15) is at (**http://www.gwu.edu/~clade/faculty/lipscomb/Cladistics.pdf**).

The Compleat Cladist: A Primer of Phylogenetic Procedures, by E.O. Wiley, D. Siegel-Causey, D.R. Brooks, and V.A. Funk, University of Kansas Museum Of Natural History, Special Publication (171 pages) (24) (**http://www.nhm.ukans.edu/CompleatCladist.pdf**).

CONCLUSIONS

Given this huge variety of methods for computing phylogenies, how can the biologist determine what is the correct method for analyzing a given data set? Published papers that attempt to address phylogenetic issues generally make use of many different algorithms and data sets in order to support their conclusions. In some cases different methods of analysis can work synergistically. Neighbor Joining methods generally produce just one tree, which can help to validate a parsimony or maximum likelihood method if that tree is also present among the possible choices. In addition, bootstrapping methods such as resampling or recalculating with random sequences deleted from the sample set can give an indication of the robustness of a given conclusion (5).

Making phylogenetic analyses of protein families is one of the most challenging and interesting tasks in bioinformatics. Because this type of analysis is rarely done, it can almost always make a valuable contribution to a research area. For example, new genes are being discovered on a daily basis in the Genome Project sequencing efforts, but little is being done to establish the evolutionary relationships between these genes and their homologs. A biologist with an interest in phylogenetic analysis could make a significant contribution to the annotation of these sequences.

REFERENCES

1. **Brodsky, L.I., A.L. Drachev, A.M. Leontovich, and S.I. Feranchuk.** 1993. A novel method of multiple alignment of biopolymer sequences. Biosystems *30*:65-79.
2. **Brodskii, L.I., V.V. Ivanov, I.aL. Kalaidzidis, A.M. Leontovich, V.K. Nikolaev, S.I. Feranchuk, and V.A. Drachev.** 1995. GeneBee-NET: An Internet based server for biopolymer structure analysis. Biokhimiia *60*:1221-1230.
3. **Dayhoff, M.O., R.M. Schwartz, and B.C. Orcutt.** 1978. A model of evolutionary change in proteins, matrixes for detecting distant relationships, p. 345-358. *In* M.O. Dayhoff (Ed.), Atlas of Protein Sequence and Structure, Vol. 5. National Biomedical Research Foundation, Washington, DC.
4. **Farris, J.S.** 1983. The logical basis of phylogenetic analysis, pp. 1-36. *In* N.I. Platnick and V.A. Funk (Eds.), Advances in Cladistics. Columbia University Press, New York.
5. **Felsenstein, J.** 1985. Confidence limits on phylogenies: an approach using the bootstrap. Evolution *39*:783-791.
6. **Felsenstein, J.** 1989. PHYLIP—phylogeny inference package (Version 3.2). Cladistics *5*:164-166.
7. **Golding, B.** 1996. Department of Biology, McMaster University, Hamilton, Ontario, Canada. Online course notes for Biology 3J03, Population Genetics; Biology 4DD3, Molecular Evolution; and Biology 721, Elementary Sequence Analysis (**http://helix.biology.mcmaster.ca/courses.html**).
8. **Hendy, M.D. and D. Penny.** 1982. Branch and bound algorithms to determine minimal evolutionary trees. Math. Biosci. *59*:277-290.
9. **Hennig, W.** 1950. Grundzüge einer Theorie der Phylogenetischen Systematik. Deutscher Zentralverlag, Berlin.
10. **Hennig, W.** 1966. Phylogenetic Systematics. University of Illinois Press, Urbana.
11. **Hinkle, G.** 1997. University of Massachusetts, Dartmouth, MA, USA. Online course notes for BIO 510; Molecular Phylogenetics and Systematics (**http://ghinkle.bio.umassd.edu/MPS/mps1.html**).
12. **Jin, L. and M. Nei.** 1990. Limitations of the evolutionary parsimony method of phylogenetic analysis. Mol. Biol. Evol. *7*:82-102.
13. **Jukes, T.H. and C.R. Cantor.** 1969. Evolution of protein molecules, p. 21-132. *In* H.N. Muntu (Ed.), Mammalian Protein Metabolism, Vol. III. Academic Press, New York.
14. **Kimura, M.** 1980. A simple method for estimating evolutionary rates of base substitutions through comparative studies of nucleotide sequences. J. Mol. Evol. *16*:111-120.
15. **Lipscomb, D.** 1998. Basics of cladistic analysis. George Washington University (**http://www.gwu.edu/~clade/faculty/lipscomb/Cladistics.pdf**).
16. **Maddison, W.P. and D.R. Maddison.** 1989. Interactive analysis of phylogeny and character evolution using the computer program MacClade. Folia Primatol. (Basel) *53*:190-202.
17. **Saitou, N. and M. Nei.** 1987. The neighbor-joining method: a new method for reconstructing phylogenetic trees. Mol. Biol. Evol. *4*:406-425.
18. **Saitou, N. and T. Imanishi.** 1989. Relative efficiencies of the Fitch-Margoliash, maximum-parsimony, maximum-likelihood, minimum-evolution, and neighbour-joining methods of phylogenetic tree construction in obtaining the correct tree. Mol. Biol. Evol. *6*:514-525.
19. **Sneath, P.H.A. and R.R. Sokal.** 1973. Numerical Taxonomy: The Principles and Practice of Numerical Classification. W.H. Freeman and Company, San Francisco, p. 230-234.
20. **Swofford, D.L.** 1993. Phylogenetic Analysis Using Parsimony PAUP, Version 3.1.1. Illinois Natural History Survey, Champaign, IL.
21. **Swofford, D.L. and G.J. Olsen.** 1990. Phylogeny reconstruction, p. 411. *In* D.M. Hillis and C. Moritz (Eds.), Molecular Systematics. Sinauer Associates, Sunderland, MA.
22. **Tajima, F.** 1993. Unbiased estimation of evolutionary distance between nucleotide sequences. Mol. Biol. Evol. *10*:677-688.
23. **Weston, P.H. and M.D. Crisp.** 1998. Introduction to Phylogenetic Systematics. Society of Australian Systematic Biologists Web site (**http://www.science.uts.edu.au/sasb/WestonCrisp.html**).
24. **Wiley, E.O., D. Siegel-Causey, D.R. Brooks, and V.A. Funk.** 1991. The Compleat Cladist: A Primer of Phylogenetic Procedures. The University of Kansas Museum of Natural History Special Publication, No. 19. Lawrence, KS.
25. **Zuckerkandl, E. and L. Pauling.** 1965. Molecules as documents of evolutionary history. J. Theor. Biol. *8*:357-366.

Section III

Appendices

Glossary

α-helix (alpha helix)

The most common 3-dimensonal secondary structure for polypeptide chains (proteins), determined by Linus Pauling in 1951. It looks like a spiral staircase in which the steps are formed by individual amino acids spaced at intervals of 1.5 Å, with 3.6 amino acids per turn. The helix is held together by hydrogen bonds between the carbony group (COOH) of one amino acid residue and the imino group (NH) of the residue 4 positions further down the chain.

accession number

A unique number assigned to a nucleotide, protein, structure, or genome record by a sequence database builder.

algorithm

A step-by-step method for solving a computational problem.

alignment

A one-to-one matching of two sequences, so that each character in a pair of sequences is associated with a single character of the other sequence or with a gap. Alignments are often displayed as two rows with a third row in between indicating levels of similarity. For example:

```
GCT---GTCTGAACCCAACCAGACGGAGAATGA
:::   ::: :: : :   :::  :::::: ::
GCTCCTGTCGGACCTCCTGCAGGGGGAGAACGA
```

alignment score

The measure of the similarity between two aligned sequence regions, calculated by summing up the scores for each aligned residue pair and subtracting gap penalties.

alleles

Alternate forms of a gene that occur at the same locus (see polymorphism). All of the variant forms of a gene that are found in a population.

annotation

The descriptive text that accompanies a sequence in a database record.

assembly

The process of aligning and building a consensus (contig) from overlapping short sequence reads determined by DNA sequencing.

β-pleated sheet (β-sheet, beta sheet)

A protein secondary structure in which two or more extended polypeptide chains line up in parallel to form a planar array that is held together by interchain hydrogen bonds. The pleats are formed by the angles of bonds between amino acids in the polypeptide chains.

bioinformatics
The use of computers for the acquisition, management, and analysis of biological information.

BLAST (Basic Local Alignment Search Tool)
A fast heuristic database similarity search tool developed by Altschul, Gish, Miller, Myers, and Lipman at the National Center for Biotechnology Information (NCBI) that allows the entire world to search query sequences against the GenBank® database over the Web. BLAST is able to detect relationships among sequences that share only isolated regions of similarity. BLAST software and source code is also available for UNIX® computers for free from the NCBI. Variants of the BLAST program include BLASTN (DNA query versus DNA database), BLASTP (protein query versus protein database), BLASTX (translated DNA query versus protein database), TBLASTN (protein query versus translated DNA database), and TBLASTX (translated DNA query versus translated DNA database).

BLOCKS
Blocks is an online database of multiply aligned ungapped segments corresponding to the most highly conserved regions of proteins documented in the PROSITE Database. Developed by S. Henikoff and J.G. Henikoff.

BLOSUM matrix
An amino acid similarity scoring matrix developed by Henikoff and Henikoff in 1992 from about 2000 blocks of aligned sequence segments characterizing more than 500 groups of related proteins. *[Henikoff, S. and J.G. Henikoff. 1992. Amino acid substitution matrices from protein blocks. Proc. Natl. Acad. Sci. USA 89:10915-10919.]*

Boolean search terms
The logical terms AND, OR, and NOT, which are used to make database searches more precise.

Central Dogma of Molecular Biology
DNA is transcribed into RNA, which is translated into protein (as proposed by Francis Crick in 1957).

cDNA
Complementary DNA—a piece of DNA copied in vitro from mRNA by a reverse transcriptase enzyme.

cladistics (phylogenetic systematics)
The method of systematics classification that groups organisms on the basis of shared derived characters.

cladogram
A branching hierarchical tree of evolutionary relationships built using cladistic analysis methods.

coding sequence
The portion of a gene that is transcribed into mRNA and translated into protein.

codon
A linear group of three nucleotides on a mRNA segment that codes for one of the 20 amino acids (see genetic code).

consensus sequence
A single DNA or protein sequence that is computed from a multiple alignment. Discrepancies between the sequences can be resolved by simple majority rule at each position, or ambiguity characters can be introduced.

conserved sequence
A base sequence in a nucleic acid (or an amino acid sequence in a protein) that has remained essentially unchanged throughout evolution.

contig
A set of overlapping sequence fragments that represent a large piece of DNA, usually a genomic region from a particular chromosome.

distance methods
A method of calculating a single measure of the amount of evolutionary change between two genomes since divergence from a common ancestor.

divergence
The gradual acquisition of dissimilar characters by related organisms over time as two taxa move away from a common point of origin (see sequence divergence).

domain
A discrete portion of a protein with its own function. The combination of domains in a single protein determines its overall function.

dot plot
A tool for the visual comparison of identity or similarity between two sequences. The two sequences are arranged along the axes of a simple graph, and a dot is placed at every point where the two sequences are identical (or similar). A diagonal stretch of dots will indicate regions where the two sequences are similar.

dynamic programming
A recursive algorithm used for finding shortest paths in graphs and in many other optimization problems, which is particularly useful in the alignment of pairs of sequences. The problem is broken down into a series of small subproblems that amount to choices between matches, mismatches, and gap insertions. When all subproblems are solved, the overall answer is determined.

EMBL (European Molecular Biology Laboratory)
The European branch of the three-part International Nucleotide Sequence Database Collaboration (together with GenBank and DNA DataBase of Japan

[DDBJ]), which maintains the EMBL Data Library (a repository of all public DNA and protein sequence data). Each of the three groups collects a portion of the total sequence data reported worldwide, and all new and updated database entries are exchanged between the groups on a daily basis. However, database files obtained from EMBL are in a different format than those obtained from GenBank. The EMBL, established in 1974, is supported by 14 European countries and Israel. Like the NCBI, the EMBL also provides extensive bioinformatics tools.

enhancer
A regulatory DNA sequence that increases transcription of a gene. An enhancer can function in either orientation, and it may be located up to several thousand base pairs upstream or downstream from the gene it regulates.

ENTREZ
ENTREZ is the online search and retrieval system that integrates information from databases at NCBI. These databases include nucleotide sequences, protein sequences, macromolecular structures, whole genomes, and MEDLINE through PUBMED.

e-score (expect value)
The expect value (e) is a parameter that describes the number of hits one can expect to see just by chance when searching a database of a particular size. An e-value of 1 is equivalent to a match that would occur by chance once in a search of that database.

EST (expressed sequence tag)
A partial sequence of a cDNA clone created by obtaining the sequence from the 3′ or 5′ end of a cDNA clone.

exon
A segment of an interrupted gene (i.e., a gene that contains introns) that is represented in the mature mRNA product—the portions of an mRNA that is left after all introns are spliced out.

FASTA
A fast heuristic sequence similarity search program developed by Pearson and Lipman. FASTA searches for local regions of similarity between sequences and is tolerant of gaps. The related programs TFASTA compares a protein query sequence to a DNA databank translated in all six reading frames, and TFASTX compares a protein sequence to a DNA databank translated in all reading frames taking frameshifts into account.

FASTA format
A simple universal text format for storing DNA and protein sequences. The sequence begins with a ">" character, followed by a single-line description (or header), followed by lines of sequence data.

gap
A space inserted into a sequence to improve its alignment with another sequence.

gap creation penalty
The cost of inserting a new gap in a sequence when creating an alignment and calculating its score.

gap extension penalty
The cost of extending an existing gap by one residue in an alignment.

GCG
A suite of software programs, also known as the Wisconsin Package, that provides a very wide range of tools for DNA and protein sequence analysis. This is commercial software developed by the Genetics Computer Group, a division of Oxford Molecular Group.

GenBank®
A repository of all public DNA and protein sequence data. GenBank is the U.S. branch of the three-part International Nucleotide Sequence Database Collaboration (together with EMBL and DDBJ). GenBank is currently administered by the National Center for Biotechnology Information, National Library of Medicine, Bethesda, MD, a division of the U.S. National Institutes of Health (NIH).

gene
A segment of DNA (a locus on a chromosome) that serves as the basic unit of biological inheritance. It includes a region that is transcribed into RNA as well as flanking regulatory sequences.

gene family
A group of closely related genes that make similar protein products.

genetic code
The correspondence between three-base DNA codons and amino acids that directs the translation of mRNA into protein. There is one standard genetic code for all eukaryotes, but some prokaryotes and subcellular organelles use variant codes.

genome
All of the genetic material in a cell or an organism.

genome project
The research and technology development effort aimed at mapping and sequencing the entire genome of humans and other organisms.

GenPept
A comprehensive protein database that contains all of the translated coding regions of GenBank sequences.

global alignment
A complete end-to-end alignment of two sequences. This can often be misleading if the two sequences are of different length or only share a limited region of similarity.

helix-turn-helix
A protein secondary structure found in many DNA binding proteins. Two adjacent α-helixes are oriented at right angles two each other.

heuristic
A computational method based on a process of successive approximations. Heuristic methods are much faster, but may miss some solutions to a problem that would be found using more laborious rigorous computational methods.

HMM (Hidden Markov Model)
A statistical model of the consensus sequence of a sequence family (i.e., protein domain). HMMs are based on probability theory—they are trained using a set of sequences that are known to be part of a family (a multiple alignment), and then they can be applied on a large-scale to search databases for other members of the family.

homologs
Sequences that are similar due to their evolution from a common ancestor.

homology
Similarity between two sequences due to their evolution from a common ancestor.

HSP (high scoring segment pair)
An alignment of two sequence regions where no gaps have been inserted and with a similarity score higher than a threshold value.

identity (see sequence identity)

informatics
The study of the application of computer and statistical techniques to the management of information. In genome projects, informatics includes the development of methods to search databases quickly, to analyze DNA sequence information, and to predict protein sequence, structure, and biological function from DNA sequence data.

init1
The alignment score derived by FASTA from the best initial alignment found between the query sequence and a database sequence in the first phase of a search.

initn
The alignment score calculated in the second phase of a FASTA search, after the shorter alignments found in the first phase of the search are joined together.

intron (intervening sequence)
A segment of DNA that is transcribed, but removed from the mRNA by a splicing reaction before translation into protein occurs.

ktup
The size of the smallest piece of the sequence that FASTA looks at when it does a search. So with ktup = 2 (the default), FASTA only looks at parts of the alignment where there are two identical amino acids in a row.

local alignment
An alignment of the region (or regions) of highest similarity between two sequences.

low complexity sequence
A DNA or protein sequence that is composed of just one or a few repeating symbols. Microsatellites and GC-rich regions are examples of DNA sequences, a glycine-rich region is a protein example.

MEDLINE (PUBMED)
The U.S. National Library of Medicine's bibliographic database covering the fields of medicine, nursing, dentistry, veterinary medicine, and the biological sciences. The MEDLINE file contains bibliographic citations and author abstracts from approximately 3900 current biomedical journals published in the United States and 70 foreign countries. PUBMED is a Web-based search tool for MEDLINE.

microsatellite
A form of repetitive or low complexity DNA that is composed of a short sequence (1–15 bp in length) that is tandemly repeated many times. This is often a hotspot for mutations.

mismatch
In an alignment, two corresponding symbols that are not the same.

motif
A region within a group of related protein or DNA sequences that is evolutionarily conserved—presumably due to its functional importance.

mRNA (messenger RNA)
The RNA molecules that are synthesized from a DNA template (transcribed) and then serve as a template for the synthesis of protein (translation). In prokaryotes, transcription and translation occur simultaneously, but in eukaryotes mRNAs are transported from the nucleus to the cytoplasm and processed before translation occurs.

mutation
A change in DNA sequence.

NCBI (National Center for Biotechnology Information)
A branch of the U.S. National Library of Medicine, which is part of the NIH. The NCBI is the home of GenBank, BLAST, MEDLINE/PUBMED, and ENTREZ.

noncoding sequence
A region of DNA that is not translated into protein. Some noncoding sequences are regulatory portions of genes, others may serve structural purposes (telomeres and centromeres), and others have no known function.

OMIM (Online Mendelian Inheritance in Man)
An online database of human genes and genetic disorders authored and edited by Dr. Victor A. McKusick. The database contains textual information, pictures, and reference information. It also contains copious links to NCBI's ENTREZ database of MEDLINE articles and sequence information.

opt
The alignment score derived from the optimized sensitive alignment derived in the third phase of a FASTA search.

ORF (open reading frame)
A region of DNA that begins with a translation start codon (ATG) and continues until a stop codon is reached—this usually is understood to imply a putative protein coding region of DNA or an exon.

orthologs
Similar genes or proteins (homologs) that perform identical functions in different species—identical genes from different species.

PAM 250 matrix
PAM is the Percent Accepted Mutations also known as a mutation probability, i.e., the probability that any amino acid will change to any other amino acid; extrapolated up to the level of 250 amino acid replacements per 100 residues (Reference 3, Chapter 10).

paralogs
Similar genes or proteins (homologs) that perform different (but related) functions either within a species or in different species—members of a gene family. The line between orthologs and paralogs grows less distinct when proteins are compared between distantly related organisms—is a bacterial protein an ortholog of a human protein, which performs an identical function, if the two share only 15% sequence identity?

parsimony
The principle that the hypothesis that requires the fewest assumptions is the most likely to be true (i.e., the most defensible hypothesis).

PCR (polymerase chain reaction)
A method of repeatedly copying segments of DNA using short oligonucleotide primers (10–30 bases long) and heat stable polymerase enzymes in a cycle of heating and cooling so as to produce an exponential increase in the number of target fragments.

PFAM
An online database of protein families, multiple sequence alignments, and Hidden Markov Models covering many common protein domains, created by E.L.L. Sonnhammer, S.R. Eddy, E. Birney, A. Bateman, and R. Durbin. PFAM is a semi-automatic protein family database, which aims to be comprehensive as well as accurate.

pharmacogenomics
The use of associations between the effects of drugs and genetic markers to develop genetic tests that can be used to fine-tune patient diagnosis and treatment.

phylogenetics
The field of biology that deals with the relationships between organisms. It includes the discovery of these relationships and the study of the causes behind this pattern.

phylogeny
The evolutionary history of an organism as it is traced back, connecting through shared ancestors to lineages of other organisms.

polymorphism
A difference in DNA sequence at a particular locus.

primer
A short DNA (or RNA) fragment that can anneal to a single-stranded template DNA to form a starting point for DNA polymerase to extend a new DNA strand complementary to the template, forming a duplex DNA molecule.

ProDom
An online protein domain database created by an automatic compilation of homologous domains from all known protein sequences (SWISS-PROT + TREMBL + TREMBL updates) using recursive PSI-BLAST searches.

promoter
A region of DNA that extends 150 to 300 bp upstream from the transcription start site of a gene that contains binding sites for RNA polymerase and regulatory DNA binding proteins.

PROSITE
PROSITE is the most authoritative database of protein families and domains. It consists of biologically significant sites, patterns, and profiles compiled by expert biologists. Created and maintained by Amos Bairoch and colleagues at the Swiss Institute of Bioinformatics.

protein family
Most proteins can be grouped, on the basis of similarities in their sequences, into a limited number of families. Proteins or protein domains belonging to a particular family generally share functional attributes and are derived from a common ancestor.

proteomics
The simultaneous investigation of all of the proteins in a cell or organism.

PUBMED
PUBMED is a Web-based search tool for MEDLINE at the NCBI Web site.

query
A word or number used as the basis for a database search.

scoring matrix (substitution matrix)
A table that assigns a value to every possible amino acid (or nucleotide) pair. This table is used when calculating alignment scores.

selectivity
The ability of a program to discriminate between true matches and matches occurring by chance alone. A decrease in selectivity results in more false positives being reported.

sensitivity
In this context, the ability of a program to identify weak but biologically significant sequence similarity.

sequence divergence
Changes in the DNA or protein sequences of homologus genes in different species due to the independent accumulation of mutations and the action of natural selection since these species shared a common ancestor.

sequence identity
The percentage of residues that are identical between two aligned sequences.

sequence similarity
The percentage of amino acid residues similar between two aligned protein sequences. Usually calculated by setting a threshold score from a scoring matrix to distinguish similar from not similar and then counting the percentage of residues that are above this threshold.

Sequence homology
The relationship between two sequences that have descended from a common ancestor.

Sequence tagged site (STS)
A short (200–500 bp) DNA sequence that has a single occurrence in a genome and whose location and base sequence are known.

shotgun method
A sequencing method that involves randomly sequencing tiny cloned pieces of the genome with no foreknowledge of where on a chromosome the piece originally came from.

signal sequence
A 16 to 30 amino acid sequence, located at the amino terminal (N terminal) end of a secreted polypeptide, that serves as a routing label to direct the protein to the appropriate subcellular compartment. The signal sequence is removed during posttranslational processing.

significance
A statistical term used to define the likelihood of a particular result being produced by chance. Significance values for sequence similarity searches are expressed as probabilites (*P* values or e-values) so that a value of 0.05 represents one chance in twenty that a given result is due to chance.

similarity (see sequence similarity)

Smith-Waterman algorithm
A rigorous dynamic programming method for deriving the optimal local alignment between the best matching regions of two sequences. It can be used to compare a single sequence to all of the sequences in an entire database to determine the best matches, but this is a very slow (but sensitive) method of similarity searching.

SRS (sequence retrieval system)
A database indexing and searching system created by Dr. Thure Etzold to identify sequences in sequence databases and retrieve them over the World Wide Web.

substitution matrix (see scoring matrix)

SWISS-PROT
A curated protein sequence database, which provides a high level of annotations, a minimal level of redundancy, and a high level of integration with other databases. SWISS-PROT contains only those protein sequences that have been experimentally verified in some way—none of these hypothetical proteins of unknown function.

systematics
The process of classification of organisms into a formal hierarchical system of groups (taxa). This is done through a process of reconstructing a single phylogenetic tree for all forms of life, which uncovers the historical pattern of events that led to the current distribution and diversity of life.

taxa
A named group of related organisms identified by sytematics.

threading
A method of computing the 3-dimensional structure of a protein from its sequence by comparison with a homologus protein of known structure.

transcription
Synthesis of RNA on a DNA template by RNA polymerase enzyme.

translation
Synthesis of protein on an mRNA template by the ribosome complex.

TrEMBL (translations of EMBL)
A database supplement to SWISS-PROT that contains all the translations of EMBL nucleotide sequence entries not yet integrated into SWISS-PROT.

UniGene
An online database (at NCBI) of clustered GenBank and EST sequences for human, mouse, and rat. Each UniGene cluster contains sequences that represent a unique gene as well as related information such as the tissue types in which the gene has been expressed and the map location.

Bioinformatics Software

Function	Name	Platform	Developer/Provider
Multiple Alignment	CLUSTAL	Web/Macintosh®/Windows®/ UNIX®/VMS® **http://transfac.gbf.de/dbsearch/ clustalw.html** **http://www2.ebi.ac.uk/clustalw/**	European Molecular Biology Laboratory (EMBL)/European Bioinformatics Institute (EBI) (free) Julie Thompson, Toby Gibson, Des Higgins
	MBSALIGNER (CLUSTALW)	Web **http://mbshortcuts.com/mbsalign/**	Molecular Biology Shortcuts Pierre Rodrigues and Julie Thompson
	PILEUP (GCG)	UNIX/VMS	Oxford Molecular Group
	MACAW	Macintosh/Windows	National Center for Biotechnology Information (NCBI) (free)
	SeqVu (editor only)	Macintosh	Garvan Institute ($10 shareware)
	SeAl (editor only)	Macintosh	Andrew Rambaut (free)
	GeneDoc	Windows	Karl Nicholas

Function	Name	Platform	Developer/Provider
	CINEMA (editor only)	Web **http://www.biochem.ucl.ac.uk/bsm/dbbrowser/CINEMA2.1/**	University College London (free)
Similarity Searching	BLAST	Web/UNIX **http://www.ncbi.nlm.nih.gov/BLAST/**	NCBI (free)
	WU-BLAST	Web/UNIX **http://www2.ebi.ac.uk/blast2/**	Washington University, St. Louis Warren Gish (free)
	FASTA	Web/UNIX/VMS Macintosh/Windows **http://www2.ebi.ac.uk/fasta3/**	W.R. Pearson (free)
	GeneMatcher (Smith-Waterman)	Web (Paracel Fast Data Finder) **http://www.ch.embnet.org/software/FDF_form.html**	EMBnet-Swiss Institute of Bioinformatics
	DeCypher (Smith-Waterman)	Web (Field Programmable Gate Array—demo server)**http://timelogic.com/index_by_algo.htm**	TimeLogic Corporation (free demo)
	GenWeb (Smith-Waterman)	Web (Bioccelerator) **http://genome.dkfz-heidelberg.de/cgi-bin/genweb/main.cgi**	EBI/Compugen (free)

Function	Name	Platform	Developer/Provider
	INCA	Web **http://itsa.ucsf.edu/~gram/home/inca/**	Richard C. Graul (free)
Plasmid Drawing	Plasmid Artist	Macintosh	GeneSystems Computer Software
	Gene Construction Kit	Macintosh	Textco
	Plasmid	Windows	Redasoft
	Plasmid PREMIER	Windows	PREMIER Biosoft International
Fragment Assembly	Sequencher	Macintosh/Windows	GeneCodes
	SeqMerge	UNIX (GCG)	Oxford Molecular Group
	Assembly Lign	Macintosh	Oxford Molecular Group
	PHRAP	UNIX	Phil Green (free/licenses)
	Assembler	UNIX/source code	TIGR
PCR Primer Design	Primers!	Web **http://www.williamstone.com/primers/index.html**	Williamstone Enterprises (free)

Function	Name	Platform	Developer/Provider
	Primer3	Web **http://www-genome.wi.mit.edu/cgi-bin/primer/primer3_www.cgi**	Whitehead Institute/MIT Steve Rozen, Helen J. Skaletsky (free)
	GeneFisher	Web **http://bibiserv.techfak.uni-bielefeld.de/genefisher/**	University of Bielefeld (free)
	Web Primer	Web **http://genome-www2.stanford.edu/cgi-bin/SGD/web-primer**	Stanford Genomic Resources (free)
	Amplify	Macintosh **http://www.wisc.edu/genetics/CATG/amplify/index.html**	Bill Engles, University of Wisconsin (free)
	xprimer	Web **http://alces.med.umn.edu/webprimers.html**	University of Minnesota Medical School (free)
	DOPE	Web **http://dope.interactiva.de/dopebin/cgibin/primer_01/primer_01**	INTERACTIVA The Virtual Laboratory
	Right Primer	Macintosh	BioDisk Software
	Oligo	Macintosh/Windows	Molecular Biology Insights

Function	Name	Platform	Developer/Provider
	Primer Premier	Windows	PREMIER Biosoft International
Molecular Structure Viewers	RasMol	Macintosh/Windows	Roger Sayle (free) Glaxo Wellcome R & D
	Chime	Windows/Macintosh/ SGI	MDL Information System (free)
	MolView	Macintosh	Thomas J. Smith (free)
Phylogenetics	PHYLIP	Web/UNIX/VMS/Macintosh/ Windows/source code **http://bioweb.pasteur.fr/seqanal/ phylogeny/phylip-uk.html**	Joseph Felsenstein (free)
	PAUP	Macintosh/Windows/UNIX/VMS	Sinauer Associates, David Swofford
	MacClade	Macintosh	Sinauer Associates, Wayne P. Maddison and David R. Maddison
	MEGA	DOS/Windows	Sudhir Kumar, Koichiro Tamura, and Masatoshi Nei (shareware $15)
	DAMBE	Windows95	Xuhua Xia (free) University of Hong Kong

Function	Name	Platform	Developer/Provider
	LVB	Web/Macintosh/Windows/OS2/ UNIX/source code **http://bioweb.pasteur.fr/seqanal/ interfaces/lvb.html**	Daniel Barker (free)
Multifunction Applications	GCG	UNIX/VMS/SGI	Oxford Molecular Group
	MacVector	Macintosh	Oxford Molecular Group
	OMIGA	Windows	Oxford Molecular Group
	Vector NTI	Macintosh/Windows	InforMax
	Gene Inspector	Macintosh	Textco
	GeneTool/PepTool	Macintosh/Windows	BioTools
	DNA Strider	Macintosh	Christian Marck ($200), Centre d'Etudes de Saclay, France
	GeneJockey	Macintosh	BioSoft
	DNAsis	Macintosh/Windows	Hitachi Software Engineering
	Lasergene	Macintosh/Windows	DNASTAR

INDEX

D

E

R

S

T

U

V

W

X